BUILDING SMALL STEAM LOCOMOTIVES

A Practical Guide to Making Engines for the Garden Gauges

ONLY

BUILDING SMALL STEAM LOCOMOTIVES

A Practical Guide to Making Engines for the Garden Gauges

PETER JONES

THE CROWOOD PRESS

First published in 2008 by
The Crowood Press Ltd
Ramsbury, Marlborough
Wiltshire SN8 2HR

www.crowood.com

British Library Cataloguing-in-Publication Data
A catalogue record for this book is available from the British Library.

ISBN 978 1 8479 029 9

Typeset by Servis Filmsetting Ltd, Stockport, Cheshire

Printed and bound in Malaysia by Times Offset (M) Sdn Bhd

Contents

Acknowledgements

It has been very gratifying to receive so much help in the preparation of this book. Various commercial concerns have been supportive and I would particularly like to thank Andrew Pullen of Aster Hobbies (UK) LLP for his generous permission in allowing some of their copyrighted images to be used. Likewise, Messrs Roundhouse have generously allowed me free use of their CAD drawings in preparation of designs. My thanks also go to Accucraft Trains, USA, for permission to use an image. You will see some photographs that Geoff Munday of LightLine has kindly provided, showing the effect of lining-out engines professionally. I would also like to thank The Crowood Press for their help and guidance over technical matters. My thanks are also due to Gareth Jones for his invariably kind assistance.

Various societies generously answered specific queries. These included the Association of 16mm Narrow Gauge Modellers, the Gauge One Model Railway Association and the London and North Western Railway Society. We are fortunate that this hobby is well provided with so much goodwill. Judy Jones has my thanks for the original line drawings of some of my designs, and I also offer thanks to Denys Bassett-Jones for his help and guidance through the maze of computer graphics. Erik-Jan Stroetinga kindly provided some valuable images of the Dutch connection to this subject. Nigel Town, of the 16mm Association, went to considerable lengths, on my behalf, to ensure that my boiler testing and insurance notes were absolutely up to date. Tony Bird and Rod Peck also must be thanked for their permission to use the Steam Motor general arrangement drawing. The main photograph on the back cover was taken at the delightful Tarren Hendre Railway, created by Jeremy Ledger and Sue Willox. Gordon McLelland kindly gave permission for me to photograph the tram engine, built by Deryck Goodall, shown on page 88.

But, above all, I am delighted to acknowledge the kind support of Harold Denyer. He prepared an entirely new set of drawings for the Gauge 1 LNWR Motor Tank, using a lifetime of experience to make it a practical working engine, as well as offering guidance about his building methods.

CHAPTER 1

Introduction

There are different ways to build a live steam locomotive for the garden gauges. In this book, we will go through them together and you can come to your own conclusions about what suits you. And yes; a complete beginner can produce a working steam engine. This book will not necessarily be a treatise on best model engineering practice. For those occasions when it isn't, I apologize in advance to experienced model engineers. The plan is to describe the techniques that may be encountered and to offer several detailed designs of varying complexity. Scattered throughout the book you will also find thumbnail sketches of further possible projects, under the heading of 'Case Notes', to illustrate particular points and to provide inspiration.

As we go through these pages together, it will become apparent that particular building methods are not set in stone. There can be an element of mixing and matching as you develop your own way of working.

The term 'garden gauges' is convenient shorthand to describe those small scales suitable for building scenic model railways in the garden, as opposed to the larger, passenger-hauling steam locomotives based around pure model engineering. Their design requirements are different. We shall mostly look at the popular scales of 16mm to the foot, G scale and Gauge 1. A brief eye will be cast over O gauge and at live steam in other scales. You will see these terms explained in the table. Names of scales are often vague and illogical. Don't try to learn them all at once. Just get used to them as you work through the book.

In Gauge 1, there is a tendency towards seeing models of standard gauge locomotives as the focus of the hobby. They may spend much of their time running on elevated tracks that are fairly functional. There is a traditional model engineering approach and the end results can be superb. Driving the models is more hands on, being helped by the controlled conditions. But there is a lighter side to this scale, which produces good, simple working engines, without too many frills. This also applies to 2.5in gauge: a magnificent scale that is something of a minority sport, but has much to commend it if space is not a problem.

On the other hand, 16mm and G scale modelling are more often about simple functioning machines that form just a small part of building a complete garden railway. Our small dragons need to be able

TABLE OF SCALES

Name	Narrow Gauge (NG)/Standard Gauge (SG)	Scale	Model Gauge	Full-Size Gauge	Notes
N gauge	SG	Tiny: various	9mm	Approx. 4ft 8.5in (SG)	A technical challenge for live steam; not for beginners
HO9	NG	3.5mm to 1ft	9mm	Approx. 1m	*ditto*
OO9	NG	4mm to 1ft	9mm	2ft to 2ft 6in	*ditto*
HO	SG	3.5mm to 1ft	16.5mm	SG	*ditto*
OO	SG	4mm to 1ft	16.5mm	Roughly SG	*See* note A
O.16.5	NG	7mm to 1ft	16.5mm	2ft to 2ft 6in	
O gauge	SG	7mm to 1ft	32mm	SG	
Gauge 1	**SG**	**10mm to 1ft**	**45mm**	**SG**	
G scale	**NG**	**Various**	**45mm**	**Variable**	***See* note B (the commercial scale of LGB)**
16mm	**NG**	**16mm to 1ft**	**32mm**	**Approx. 2ft**	***See* note C**
2.5in gauge	**SG**	**Various**	**64mm (2.5in)**	**SG**	***See* note D**
⅞in scale	NG	As its name	45mm	15in	*See* note E

Notes

Note A: OO is the smallest commercial scale for ready-to-run locomotives (by Hornby), but I would suggest that it is still too small for a novice loco-builder.

Note B: G scale was originally a proportion of 1:22.5 running on 45mm track – representing a typical European narrow gauge. However, some US manufacturers in particular now apply the term to 1:24 scale – or even as 1:29 scale to represent a standard gauge. This is at odds with the pre-existing 1:32 scale of Gauge 1.

Note C: 16mm scale (also known as SM32 – 'sixteen millimetre scale, 32 millimetre gauge') is a very active scale for loco-builders. Strictly speaking, it represents around 2ft gauge in the prototype. But there were many different gauges, with some varying by only half an inch. It is possible to get tied up in knots about this, but there is a useful convention. Many narrow gauge engines were standard classes supplied in a variety of gauges to suit particular railways. Rather than have a multiplicity of model scales, one accepts the premise that 'your' railway is a model 2ft gauge and so your engine runs on 32mm gauge track – and thus provides a recognized standard.

Note D: 2.5in gauge was originally used for small live steam locomotives to haul real people on functional (usually elevated) track. It was the smallest of the 'model engineering' scales. It would be built to a scale of 17/64 to the foot. The concept was very traditional, as was the engineering. There has recently been a revival of interest in the very old 'Gauge 3'. This is still 2.5in gauge but is all about model railways in the garden. Finally, an offshoot of this is G64. This also is 2.5in gauge, but has more of a G scale ready-to-run feel.

Note E: There is a small but growing interest in modelling very narrow gauge railways – typically 15in or 18in – that are to ⅞in scale and run on 45mm gauge track. These are really big, chunky models.

Scales marked in bold type are those of the garden gauges that this book will concentrate on.

The table is somewhat simplified. There is a plethora of variations. In particular, in the USA there is a whole raft of different names. But as a rough guide, it is often common with narrow gauge names to state the scale first, then to put in an N to indicate that it is narrow gauge, and finally to put the gauge of the prototype – often in inches. An example might be On30.

The word 'approximate' occurs frequently, as do contradictions in scale nomenclature. Adherents of particular scales can be quite forceful in their views. Sorry for the complications; but you soon get used to them.

to run unattended for long periods, over undulating track, encountering the occasional leaf, snail and questionable rail joint. It is practical matters like these that give rise to the differences in design and building methods. I cheerfully acknowledge that I am simplifying things here. There are many exceptions and grey areas. What we need is some sort of framework.

FOLLOWING A FRAMEWORK

A useful first step in that framework will be to agree on all of the technical terms and these will be explained in the next chapter, 'The Naming of Parts'. It won't make exciting reading, but may avoid confusion as we proceed further together.

The initial practical step will be to take a look at the basic metalworking processes. These can have a daunting mystique to them, so the sooner we 'de-daunt' them the better. You can draw a straight line with a pencil and a ruler; you can therefore do the same thing with a scriber on metal. You may not be familiar with cutting metal but it is soon learnt – and, besides, there are a few sneaky dodges we can resort to that will overcome a lack of experience. Metalworking is partly about what you officially learn in school and partly what you learn behind the bike shed afterwards. More experienced modellers will forgive me if I devote some space to starting with the basics.

The next stage will be to look at machining processes, particularly using the lathe. This will move us up to a whole new level of expertise, although the use of a lathe isn't essential. After reading this book, you may conclude that there is a trade-off to be had between buying machine tooling and purchasing ready-made items like cylinders. If you want to build a simple 0-4-0 tank engine or two, it can make sense to buy in some components, rather than equip a machine shop and develop the expertise to use it. This brings us onto the important subject of your own motivation.

There are those amongst us for whom the challenge of scratch-building a model railway engine is the sole ambition. It may never turn a wheel, once completed. Such an ambition is markedly different to that of someone who wants to end up with a simple steam loco, with the minimum of fuss, to run on a garden railway. Have a think about what your priorities are. It is a good idea to look around at all sorts of different locomotives, at exhibitions, in magazines and in clubs. By all means talk to others, but be aware that some people tend to focus on their own world. They are so immersed in their particular interests that they cannot always take in the fact that others see things differently. This is by no means a criticism. We need a broad base to the subject and all are welcome.

Into the framework outlined above, various small steam locomotive designs will be presented. Several complete projects, with much detail, are offered as we go along. In particular, I am delighted to say that the noted loco designer and builder, Harold Denyer, has produced a fully worked-up design for a Gauge 1 locomotive – the Motor Tank – exclusively for this book. Scattered throughout these pages is also a further selection of suggested drawings and designs of mine, which are touched on briefly, in the hope of providing inspiration. These include a group of some early, crude sketches of locos I built at a tender age. I offer these in the hope that they will provide inspiration. They come from a time when everything had to be made from what was available and, I suggest, still have a naive charm.

WHY?

Why do we want to build a steam locomotive anyway? There are all sorts of answers to this seemingly innocuous question. Yes, there is the challenge of having done it. But a more prosaic reason might be that it can be cheaper to build than to buy. There is also the fact that it opens up a whole variety of designs that cannot be bought over the counter. You can build the loco that you want. It can also be for no other reason than it is enjoyable. It has to be said: there is a quiet satisfaction to be had from seeing one's own creation, in steam and running.

We shall also look at boilermaking. You may well feel that it is easier to buy in a ready-made boiler if your needs are modest. Boilers range from a simple can shape (hereinafter referred to as 'pot boilers'), heated from underneath, or they can have a

140
120
100
80
60
40
20
0
Bert is a useful driver that comes in various scales. Feel free to copy him, cut him out and glue him to a piece of backing card. If you are working in a new scale, he will help you keep a sense of perspective and avoid any obvious errors in your own loco designs.
If you are building 2.5" gauge engines, there are several scales you may be using. The LGB Bert or the 16mm one should be roughly right for your needs.
Keep him in your wallet. He can be a good friend.
0 Gauge
Gauge 1
G Scale
16mm Scale
7/8" Scale

The magic of live steam. A vintage model of an LMS 4-4-0, in 2.5in gauge.

tube running through for a burning gas jet to boil the water. Boilers can be of more complex design, but we will not look in great depth at coal-fired boilers. They are the pinnacle of loco-building and are very satisfying to manage, but they need fairly constant attention when running. This may not suit your need for a small engine that has to look after itself for half an hour on a ground-level layout. Its building also calls for work at a higher skill level.

MOTIVATION

This brings us neatly to something that is rarely written about, but which I think we should discuss

16mm narrow gauge tends to place a steam engine as part of an overall layout.

right at the start. I invite you to think about your motivation. There are those fortunate people who can concentrate on a project that goes on for months – even years. In their single-mindedness, they swerve neither to left nor right. Lesser mortals run the risk of starting a complex project that will never come to completion. Better a simple locomotive happily running round your layout, than an unfinished project eternally gathering dust on a shelf.

Metalsmithing does not call for black magic. There are techniques that can be learned. However, what you will notice at first is that the processes seem slow. If you have only worked in wood or plastic until now, some jobs like filing and sawing metal can seem endless. Fortunately, after a while, you get attuned to this pace and it becomes quite normal. But it is no use me pretending that you can build your first steam engine in a week. It is an ongoing thing. Concentrate on the job in hand and slowly work through all the steps. My advice is, as ever, work slowly and the job will be finished before you know it – rush it and it will take forever. This is particularly apt for building live steam locomotives.

When you have finished the engine, people will comment on how patient you must be. You will study the backs of your hands nonchalantly and bask in the admiration. But I am, like many others, not blessed with endless patience. To overcome this, I work on the task in hand until I get bored with it . . . then leave it and come back later. There is no rush. If there is a secret to loco-building it is to avoid all thought of time; just let the jobs flow gently past you at a comfortable rate.

COSTS

It is very satisfying to buy a brand new shiny steam engine and enjoy running it. If ever there was an example of immediate gratification, this is it. Buying materials for something that may get built one day seems markedly less inviting. Moreover, it can seem that there is no guarantee that the engine will work when it is finished. Rest assured, we have all shared those thoughts at some time or other. The thought of paying good money to go into entirely unknown territory can be unnerving at first. Fortunately, we can usually get round that. There are escape routes if something goes wrong. By starting off with a simple project, using ready-made components, we can be sure that the end result will be

Roundhouse Models technology can be a useful resource for a loco builder.

a nice little runner. And along the way, we will have learned a few simple skills to take forward to other projects.

It is a sad fact of life that a fully equipped workshop doesn't come cheap. However, such a thing isn't essential. If you reach the stage of building a steam engine entirely from raw materials then your building costs will remain low thereafter. Until then, you can buy in cylinders and boilers, and get by with hand tools, although these need to be of good quality. Don't begrudge buying decent hacksaw blades and a few new files. That said, there are a few sneaky ways to get a job done without expense. Instead of buying a scriber, I used an old set of darts, with tungsten tips, for many years. And there is plenty of scope for making temporary jigs out of bits of wood and nails, even though we are going to need considerable accuracy with our work. Life is easier – and cheaper – with a bit of thought and planning. You will find examples and suggestions of this, scattered throughout these pages.

THE PLACE OF WORK

I could describe a 'perfect workshop' (whatever that is) and suggest you should build it. But in the real world, we make the best of what is available. This might be a small shed, the corner of a garage, or even a hobbies room. The actual working area can be quite small, but will be more comfortable to work in if made as civilized as possible.

SAFETY

Please don't skip this section, even if much of it seems obvious. I will try to avoid describing safety measures so severe that you wouldn't want to get out of bed in the morning. That would be counterproductive. However, small-scale model engineering offers new opportunities for harming ourselves and others, most of which are avoidable. You probably won't need telling that sharp tools are good at cutting flesh. In my experience though, blunt ones are usually the cause of more accidents. Keep tools sharp and maintain a healthy fear of their sharpness. Raw pieces of metal can lacerate nicely too.

Use a circuit breaker. With mains electricity there are no alternatives. Do everything right. Adhere to Building Regulations. Replace fuses with correct values and always be unconsciously on the lookout for damaged wiring.

Soldering irons have two ends: one of them gets hot. Do not pick it up by this end. You would be surprised at how many people, when pressed, will admit to getting this wrong, the author included. Make sure the hot end can't fall onto something flammable. There are little gadgets that look like a bed spring to hold the iron when not in use. I have a length of very thick-walled steel tube, fitted near the bench. When I have finished with the iron, I post it into the top of the tube, which is so thick and chunky that it conducts away any residual heat quickly and safely.

Blowlamps are good at starting fires. In times past, we had to use fierce pump-me-up gadgets that could throw jets of flame across the room nicely if not properly used and maintained. But now we have the more respectable gas blowlamps – either with a disposable can screwed underneath, or with a hose to a bigger bottle.

Although high temperatures are reached by the flame, they are localized, so a simple hearth, described later, will contain the heat safely. But develop a healthy regard for how easy it is to start a fire. Never throw anything even warm into a waste-paper bin. If I have been sawing a piece of metal, I will leave the offcut somewhere to go absolutely cold before disposing of it. The same goes for a tool that has just been sharpened on a grinding wheel. It is a good habit to get into and to do automatically.

Keeping flammable liquids away from hot areas makes sense. If you are short of space, keep them in a stout box. Indeed, you should consider having a flammable store outdoors. Don't begrudge buying a dry-powder fire extinguisher.

Some processes can create fumes. If something says 'not to be used in an enclosed space', then, for your own sake, *don't*. Disposable masks will cope with dust and heavier particles, but some chemical reactions or some paint sprays call for properly approved masks if used indoors. On the subject of protective wear, *do* use safety goggles when carrying out anything that could remotely damage your eyes. This includes grinding, turning or power-sawing metal.

All of this sounds obvious. However, people can be really ingenious when it comes to inventing new ways of harming themselves and destroying property. So, as ever, such things are for the guidance of wise men and the protection of fools. It all comes down to common sense – and simply getting into the habit of thinking safe.

The main need is for a stout workbench at a comfortable height. I like my working area to be higher than a kitchen table: with advancing years it is kinder on the back and the eyesight. My workshop chair is therefore a high wooden stool.

A concrete garage can be prone to condensation. If you can't get around that, keep tools and metal bits inside wooden boxes when not in use – and be liberal with the WD-40 spray. How you heat a workplace is up to you. The only observation I would make is to avoid any heater that burns paraffin with a naked flame: it adds to any problem of dampness. I have also stood in the burnt-out remains of a workshop where someone had knocked over a heater and its automatic cut-out device didn't work.

But the one thing that makes all the difference is having plenty of light, which means not just a 60W bulb in the middle of the room behind your shoulder. Give yourself several power points close at hand, protected by a circuit breaker. Make all surfaces light-coloured. Working on a really good surface is somehow kinder. A good idea is to recycle pieces of white-faced kitchen top to lay on your workbench. When one gets dirty, replace it.

You *can* build a steam loco if you have just a stout wooden bench in the garden. I'm not going to pretend it is humane, but it is possible. Do what you can indoors, on a table: things like marking out and filing jobs spring to mind. You will need a stout clamp-on vice to take with you when you go outside.

Life is also easier if you are a naturally tidy person. Some of us are not . . . but wish we were.

Doing research into a project does not call for metalworking skills. Enjoy the process of learning all the ins and outs of particular locomotives and the ways of building them. At the back of the book you will find suggested further reading and lists of societies to contact. There is much information to be found on the Internet. Build up a picture of what you want to do, long before you cut metal. If I am trying out something completely new, I will often put together a card (or even metal) mock-up so that I can see how parts relate to each other – and to get a feel for the sizes of things.

One last preliminary is to make sure we learn the language, which is why the next chapter is called 'The Naming of Parts'.

CHAPTER 2

The Naming of Parts

To avoid any risk of confusion as this book unfolds and with the complete beginner in mind, this chapter will run through some of the technical terms that can be encountered when describing locomotives, both full size and models.

Starting at the bottom, wheels are defined by their diameter and the number of spokes. A pair of wheels on an axle is called a wheel set. In real life, wheels are cast in iron or steel, but have steel tyres shrunk onto the outside. These tyres have a flange that sits on the inside face of the rail. The distance between the rails is the gauge (note spelling) and the matching wheel set is referred to as being of that gauge. Cast into the centre of each driving wheel is the boss. In the case of a conventional inside framed engine, this is eccentrically shaped and from it juts a crankpin. Driving wheel sets are linked together with coupling rods. These are made from steel, often with adjustable bearings that enable them to be kept to a good working fit on the crankpins. They are lined with a bearing metal, typically gunmetal or bronze. Such things are often referred to as the 'brasses'.

Cylinders are either outside the frames, where they are visible, or tucked between them out of

Outside cylinders, spark-arresting chimney, simple buffer/coupling, narrow gauge: all will become clear.

sight. The motion to drive a crankpin is transmitted by a connecting rod, coming from one of the cylinders. If these cylinders are between the frames (an inside cylinder loco), then the driving axle has to have U-shaped bends, called cranks, allowing it to be driven round and round by an inside connecting rod. In fact, these U-shapes were rarely produced by bending, instead being mostly forged and machined. In our models we will fabricate them. The straight motion coming from a piston – the piston rod – is converted by a knuckle joint (the little end) to the up and down motion produced by the other end of the connecting rod – the big end – driving the rotating wheel.

With all these heavy rods flailing about there would be massive hammer blows upwards and downwards. To neutralize these, balance weights are used. If you look at a variety of driving wheels you will see a crescent-shaped area, called a balance weight, inside a portion of the wheel rim. This is the most obvious form of wheel balancing. Such forces in our small models will be tiny, but they are still modelled for the sake of appearance. Fortunately, wheel castings often come with correct balance weights cast in.

The cylinders let steam in at alternate ends (and exhaust out at the opposite ends), which is called 'double acting'. To control the admission and exhaust of the steam, some form of valve gear is needed. The actual valve is found in the valve chest, which is a steam-tight area adjacent to the cylinder itself. Normally one encounters two sorts of valve. The first consists of a thick, flat plate, which has a recess in it. As this slides to and fro on a flat surface, that recess connects passages cast into the steam chest. This is called a slide valve. The other option is the piston valve. The valve chest is bored out and a long, thin piston slides backwards and forwards. This piston has deep grooves cut in it and it is these that allow a free flow of steam from one opening to another. To a complete newcomer it may all seem complicated, but, I promise, you will be completely familiar with it by the end of this book. The distance a valve travels, and its timing, is controlled by the valve gear. We will look at this in depth later on.

A full-size locomotive boiler has a firebox at the back end. The heat from this is drawn through tubes in the boiler proper. At the front end is the smokebox, where the remaining gases are drawn out and exhausted up the chimney. This needs additional draught to suck those gases through. Cunningly, the exhaust steam from the cylinders is blasted upwards through the chimney, drawing the hot gas with it. When an engine is in steam but not moving, a separate jet of steam, taken from the boiler, is shot up the chimney. This is known as the blower and, logically enough, is operated by the blower valve.

The back of the firebox, where all the controls are located, is called the backhead. It is usually enclosed by the cab, although some small engines, particularly narrow gauge, don't have this luxury and are referred to as open-cabbed. The fire is

A set of part-machined cylinder castings in bronze.

burnt on a grate, but to stop ashes falling through onto the track, there is an ashpan underneath.

BOILERS

Building these will be described in detail in Chapter 8, but we can look at some of the peripherals here. Steam engines boil huge amounts of water. This is stored either in tanks bolted onto the engine itself (hence the term tank engine), or in a separate tender. Water has to be forced into the boiler against the steam pressure. This is done either by a pump or by an amazing item called an injector. This has no moving parts, but, by using cones to accelerate steam from the boiler and combining it with the feed water, it injects water against the boiler pressure. However, the injector does not lend itself to the very small steam models we are interested in, so we will only be looking at various forms of pump. A driver needs to know the pressure of the boiler and this is done with a pressure gauge. But, most of all, he needs to know the water level in the boiler. A water-level gauge is a roughly vertical glass tube, both ends of which are connected to the boiler.

Some primitive engines featured vertical boilers, occasionally with gloriously polished domes. But normally, in a horizontal boiler, the firebox would be round-topped and appear to be just a continuation of the boiler. A later development was the Belpaire firebox, which had an appearance of having raised square 'shoulders' near the top.

In models, the water is usually topped up in one of three ways. The first, particularly favoured in narrow gauge models where the boiler can be quite large, is to run the boiler until it is nearly empty and then kill the heat. As soon as the remaining pressure has been released, a filler plug is unscrewed and water is squirted in, either by a big plastic syringe or by a squeezy water bottle – the ones used to fill model aeroplane engines are excellent. Another alternative is a cheap plastic plant spray bottle, adapted to take a length of plastic tube. This method of boiler filling seems quite acceptable where an engine can run for forty to sixty minutes on one filling.

Smaller boilers might need topping up more frequently. There is a little gadget called an Enots valve, which allows a tube to be fitted pressure-tight, via a non-return valve, to the boiler. A few pumps from a plant sprayer bottle will overcome boiler pressure and the boiler is then refilled.

The Goodall valve (named after its inventor) is a particularly ingenious non-return valve in its own right. It consists of a stub of flexible tube that is slipped over an opening in a hollow brass stem. The boiler pressure normally crushes the tube against the opening and prevents water escaping, while

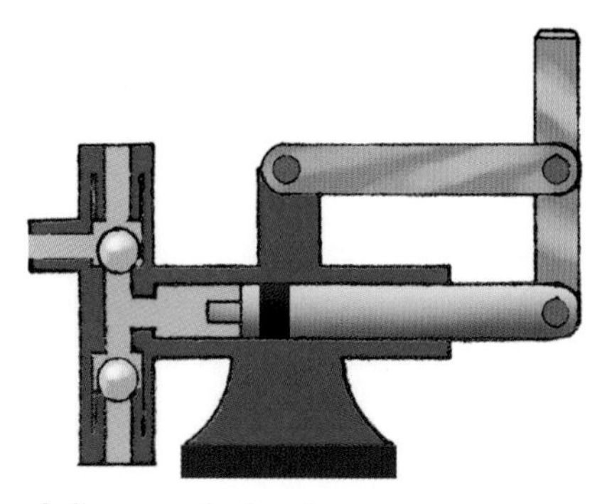

A diagram of a handpump.

greater pressure from the opposite direction forces the tube away from the stem and allows water to be pumped in.

The traditional method of filling a boiler is a hand pump with non-return valves located in a side tank or tender. The handle is waggled to and fro with a plug-in extension lever. Larger scale engines have an axle-driven pump. This can create a drag on very small engines as they run along and so its use is not so common.

Finally, I should like to mention a particularly elegant way of filling a boiler. A simple tap is fitted to the backhead. When the fire is extinguished, the boiler cools. As it does so, it creates a vacuum. By attaching a plastic tube to this tap, the other end will suck in water from a container.

FRAMES

The 'backbone' of an engine is its frames. In Britain these are normally flat plates of steel on edge (plate frames), held the correct distance apart by frame spacers. American engines usually feature frames made of more flexible steel bars. The front end will sport a buffer beam. If the engine is a self-contained tank engine, the rear will also have a buffer beam. If it is a tender loco, then the beam has no buffers and is called the drag beam.

Buffers are usually sprung and consist of two parts: the buffer head, which slides in and out of the other component, the buffer stock. Narrow gauge locomotives often have a single centre buffer/coupling assembly. These come in a multitude of patterns. Many engines will feature chain-link couplings, but we will also encounter automatic coupling in the form of knuckle couplers. I heartily

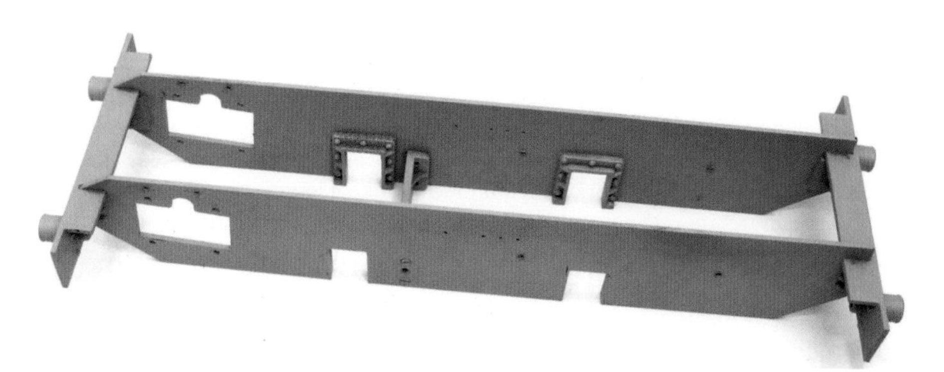

A set of frames, assembled with the buffer beams. There are cut-outs for sprung axle boxes and for inside valve chests.

American-type bar frames call for a lot of precise metalwork. Apart from the project in Chapter 4, they will not be needed in this book.

A rolling chassis with cylinders and smokebox saddle in place.

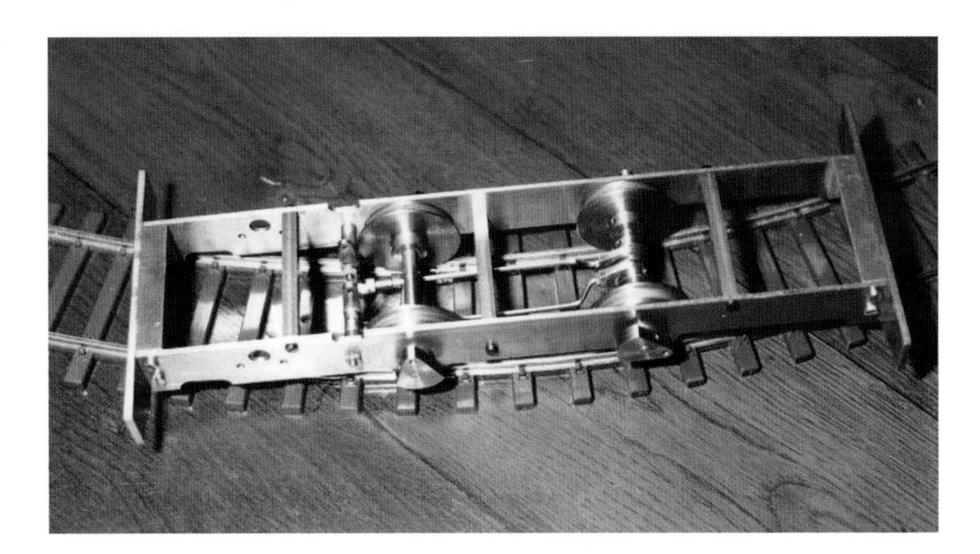

A scratch-built 16mm NG chassis. It shows how the sharpness of a curve can limit our options. Photo: Erik-Jan Stroetinga

commend the Internet as a splendid place to study further the working parts of an engine. In particular, there are animated diagrams of how valve gears work: most illuminating.

SCALES AND GAUGES

Following on from the Table in Chapter 1, which gave bare bones, we can now go further. The scale of a model is expressed either as a fraction, for example 1/24, or as an equivalent, such as 10mm to the foot. The thing that strikes you immediately is that metric and imperial scales are used together indiscriminately. This is just a fact of life that everyone gets used to. Fortunately, the cheap pocket calculator takes much of the effort out of this. If you want to build a loco in milli-cubits to the foot, then carry on. You can also buy scale

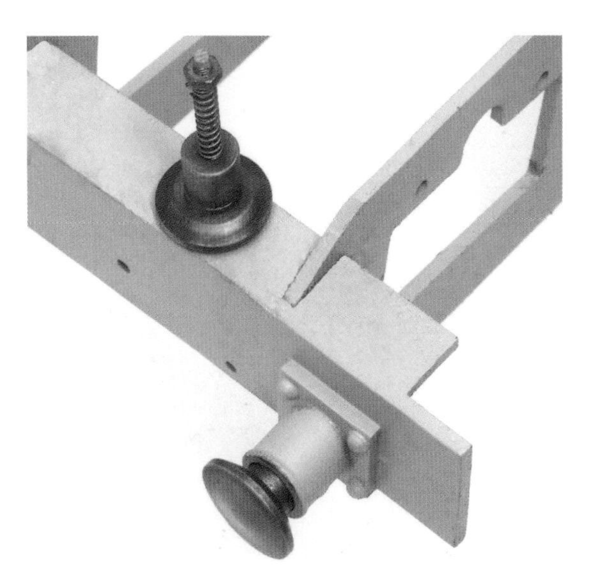

This buffer assembly consists of the fixed stock and a sprung head.

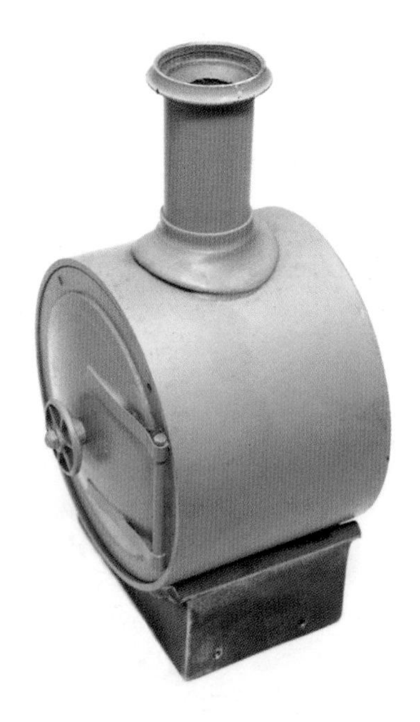

A drum-type smokebox sits on its saddle. The chimney is homemade from separate parts . . .

. . . compared to this example, which is machined from a single casting.

rulers for converting dimensions. In due course, you will get so immersed in the way scales relate to each other that it ceases to be a problem. You can also invent your own scale gauge combination. If you want to run a model of a 15in gauge steam engine on 2.5in gauge track, it will be to a huge scale. The permutations are endless and many have not been tried yet.

MODEL LOCOMOTIVE TERMS

Two frames, assembled with spacers and beams – and to which the wheels have been added – becomes the rolling chassis. A steam chassis describes when the cylinders and motion have been fitted to a rolling chassis and the assembly will run on air or steam. The term 'bodyshell' is not common in live steam because it describes a complete locomotive shape that drops over working mechanisms concealed underneath. But we may encounter it in relation to tram engines. There are various circumstances where steam engines could run through public streets and byways, where they would be subject to considerable safety restrictions. Typically, all the moving parts below the running boards would have to be hidden behind a casing. Often the prototypes would be limited to, typically, 8mph. However, to take this principle a stage further, there would be completely encased locomotives. In the early days of urban tram operation, such locomotives would haul a substantial tram-type trailer. To obtain in model

Gaskets are used, as in cars, to produce a pressure-tight joint between mating flat surfaces.

COMPUTER HELP

On the subject of converting sizes, a computer is very useful for accurately enlarging or reducing drawings. Even the simplest graphics programs have facilities for altering image size. On the subject of copyright, no one is going to complain if you rescale a drawing you have bought for your own use, but you would have to seek permission to make copies for sale, or to publish your amended drawing.

To change the scale of a drawing in, say, Photoshop Elements, a typical procedure might go like this: bring up Rulers. Crop to overall length. Note the Image size. Adjust this by the change of scale fraction you have scribbled on the back of an envelope. Print out. Fortunately, the side elevation of many small locomotives, in the garden gauges, will fit onto a sheet of A4.

If you are working off an old blueprint, you can Invert the image and then desaturate to produce a black on white plan. This can be tidied up with the Contrast and Eraser tools. The example used shows a battered portion of a huge blueprint, taken with a camera. This is reduced in size and cleaned up. Not perfect, but not rocket science to do. For smaller originals, scan them into the computer and enlarge.

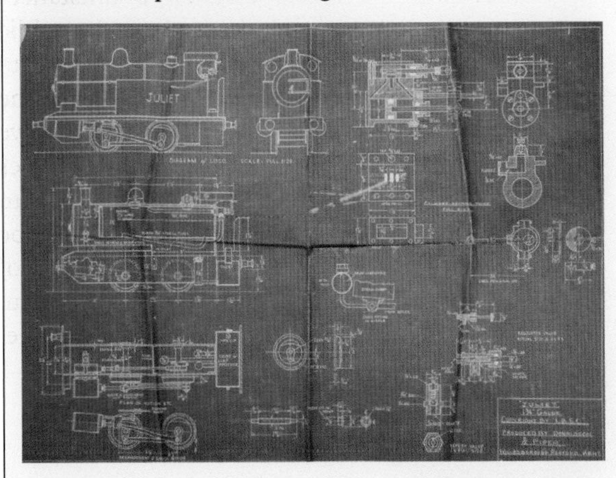

The original battered blueprint, from which I built a model many years ago.

You can also home in on any known dimensions. If you have an unknown simple side elevation of a loco, but where you have read somewhere that the wheels of the prototype were 4ft 6in diameter, with the aid of the computer you can quickly produce the drawing to the scale you are working in. If you are particularly competent with manipulating graphics, it is possible to extrapolate side and front elevations from perspective photographs. Fortunately, the need to do so is rare because so many drawings already exist.

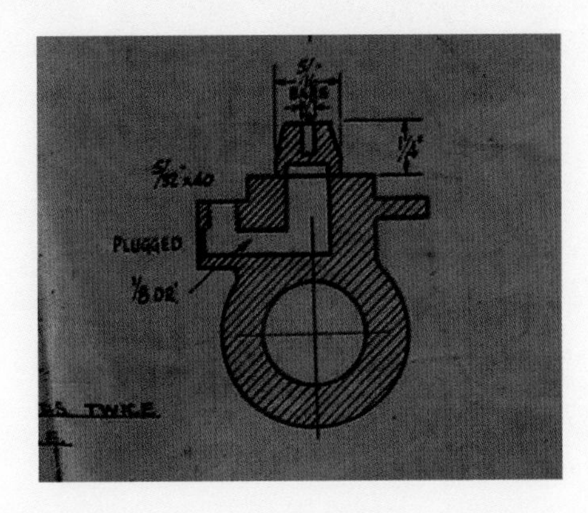

The drawing is scanned and the colours inverted.

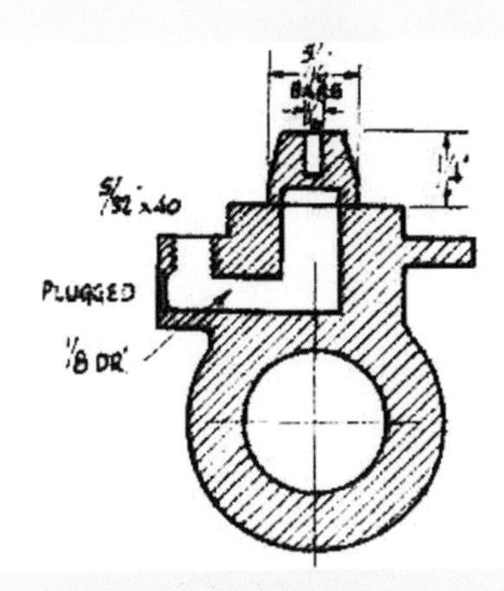

Finally, it is brought up to white with Brightness and Contrast controls. It is cleaned up with the clone tool and is now ready to be used.

form the really slow, smooth running of a tram engine, we might resort to having a cylinder driving an intermediate shaft and then a gear reduction down to the wheels. It is thus a geared engine.

The simplest way to fire a boiler is to have naked flames underneath. In the early days primitive engines were called potboilers and were externally fired. They were fairly useless in all but absolutely calm conditions. They also produced lots of flames, which damaged paintwork (and hands). However, from the 1960s onwards, a refined breed of potboiler appeared. They had various forms of shielding

A tiny steam tram engine in OO gauge, belonging to Alan Craney. The bodyshell conceals a very functional steam mechanism.

The Glyn Valley Tramway shell also conceals very functional workings. This was built by Robin Gosling some forty years before the time of writing. The geared oscillating engine still runs smoothly.

The interior of a well-used bodyshell.

STEAM MOTOR

From time to time, there have been useful little devices for sale. They usually consist of one or two small oscillating cylinders, with a gear reduction down to an axle – usually for 16mm scale in 32mm or 45mm gauge. They are intended to be a straight replacement for an electric motor and, in theory, go into any rolling chassis. Such things are a short-cut to a working steam engine. It is better to acquire a working one first and then design a locomotive around its dimensions.

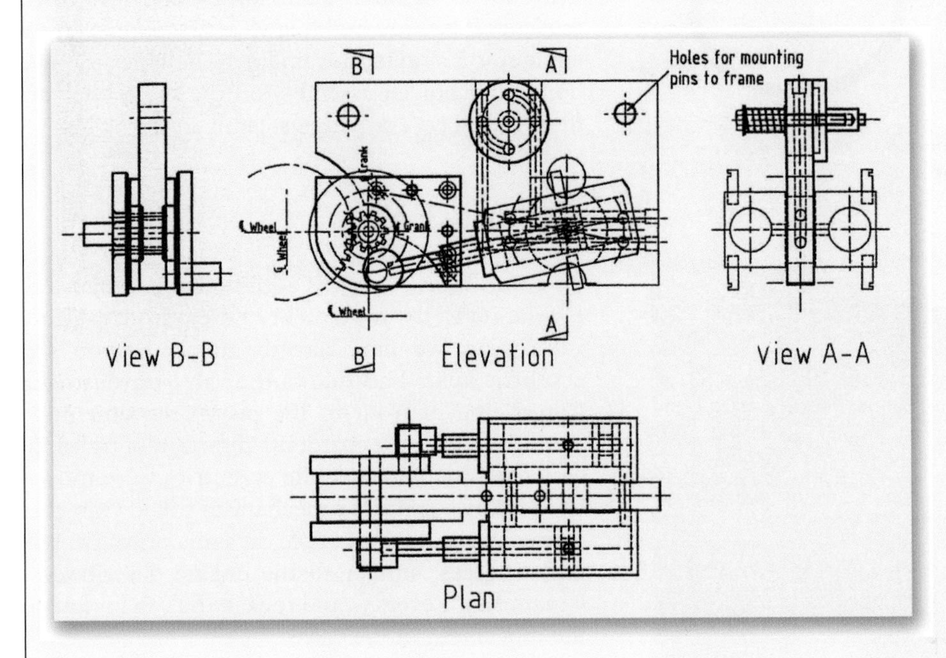

A typical steam motor. This example was developed by Tony Bird and is particularly small and neat.

around the flames, were hard soldered and ran at a higher pressure. And here tribute should be paid to Stuart Browne, of Archangel Models, whose designs transformed unreliable toys into controllable and well-engineered workhorses. This was taken further by Roundhouse Engineering, which developed an invention by pioneer Jack Wheldon. This was a form of inner wrapper that held the flames close to the lower half of a boiler, increasing its efficiency and providing considerable protection to a stable flame in windy conditions. It was called a Pooter firebox, named after Jack's locomotive that first used it (from the character Charles Pooter in the book, *Diary of a Nobody*).

In the past, an answer to protecting naked flames was to use a boiler within a boiler. Inside a cosmetic outer shell, there would be an inner boiler (Smithies boiler). The underside of the firebox is open and contains wick tubes. The flames are drawn up and along, between the boiler and outer wrapper, and out through the chimney. This needs some sort of artificial draught to assist it. When an engine like this is first being lit up, a small electric fan is placed over the chimney to suck the hot gases through. Once a modicum of steam is raised, a small jet of steam is blown up through the chimney and this sucks the hot gases with it. Akin to the full-size practice described earlier, this is about the most basic form of internal firing. It has keen support in Gauge 1 in particular and boiler design has been considerably refined to a point where very efficient and practical steam-raising is the norm.

The ultimate form of internal firing is, of course, by coal. I am going to suggest that the complex engineering involved makes this unsuitable as a first project for a beginner, although throughout this book there will be signposts that point towards finding out more. An obvious first port of call is the hobby of

Heat resistant O-rings and fibre washers have mostly replaced packed graphited yarn.

Here we have an extreme close-up of a tiny fitting, which will seat on an O-ring.

larger-scale model engineering, but I would make the point that merely scaling down large-scale traditional models is rarely satisfactory. There are all sorts of minor differences that small-scale coal firing seems to demand. It may be more successful to contact existing builders of small coal-fired engines, to get a feel for the hints and tips that can make a difference. I would also make the point that building a coal-fired loco represents a major commitment, especially if it is to be just a part of an overall garden railway.

In the narrow gauge garden layouts, commercial gas firing has come to dominate the hobby. This can be as an external flame under the boiler, together with a Pooter-type firebox, or more likely it is a jet of flame that runs through a central flue tube. It is clean, convenient and fuss-free. No blower is needed and the heat is very predictable. I will always have a soft spot for the traditional method of firing locos on methylated spirits, but have to be realistic. If you are building your first simple steam engine, it is a great advantage to be able to buy a boiler and burner assembly off the shelf, knowing that it will give consistent steaming.

VALVE GEAR

To control how steam is admitted and exhausted, in relation to the movement of the piston, requires valve gear. We have already touched upon slip eccentric gear. This allows the valve travel to slip from being in front of the piston movement, to behind it – and so control the direction of travel. In very early locomotives, the eccentrics were moved by hand by a primitive mechanism called Gab gear (this was used, for example, in some primitive box tank engines, similar to the engine described in Chapter 4). Levers would rock in the cab in time to the movement of the valves.

Steam is more efficiently used in full-size engines by varying the amount that is allowed to expand in the cylinders. For starting off, it might be admitted for 65 per cent of the stroke of the piston. This is what causes the noisy 'whumph-whumph-whumph' of a steam train pulling away. But once on the move, less steam is needed. Steam is then only admitted for a short part of the stroke and is allowed to expand more fully. To achieve this, there have been two predominant valve gears. The Stephenson gear does the job well and efficiently – and has done for a very long time. To many people though, this was improved on by Walschaerts gear, which did the job just as well and with fewer parts. In a minority are things like the Joy and Hackworth gears. They are less efficient than the principal valve gears, but can be simple to produce. In standard gauge practice there are also rotary cam gears. But in terms of simple model steam engines, we are not worried about the last percentage of efficiency. We seek models that are easy to build and run . . . and

FIRST PRINCIPLES

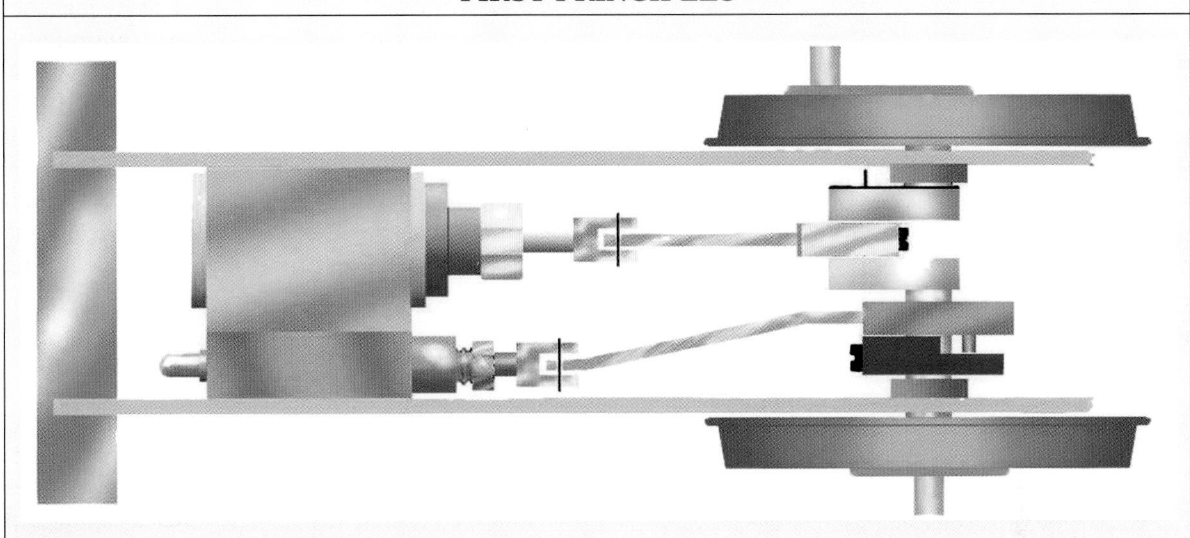

A basic single-cylinder assembly in diagrammatic form.

The drawing here is intended to show how the various working parts of a steam engine go together. If you are completely new to this subject, it would be useful if we looked at it together.

There is a front buffer beam on the left and two mainframes coming from that. Between the frames is a single slide-valve cylinder, connected to a crank axle. So this would be called an inside single-cylinder engine, operated by slip eccentric gear.

The big square block is what a cylinder would look like, viewed from above, if it were machined from a single piece of gunmetal or phosphor-bronze (alloys similar to ordinary brass but more hard-wearing). The block will be accurately machined flat on all faces and there will be an equally accurate bore, through the centre, that is at a true right angle to the ends.

The smaller rectangle next to it – the valve chest – is an enclosed space in which a little valve slides backwards and forwards, uncovering openings on the face of the cylinder block. This will be more fully explained below in a similar diagram dedicated to an actual cylinder assembly. A rod goes into the valve chest, as you can see. Occasionally you will come across a design where the rod – sometimes called a valve spindle – is made longer and projects into a little extension piece that looks like a tiny lipstick case and which gives additional support. I have shown one in the diagram, but they are not always present.

The fact that a piston, moving backwards and forwards, imparts a rotary motion to the wheels, via a cranked axle, is clear. But less obvious is the fact that a disc of metal is fitted, off-centre, onto the axle, thus making it wobble. Surrounding this 'eccentric' is a strap that is a snug but free-moving fit around it. This strap is connected to a valve rod that goes into the valve chest. Because the eccentric is tumbling, this imparts limited movement to the valve rod as it rotates.

The angle between the actions of the crank and the eccentric determine the timing of events. By rotating the eccentric on the axle (and locking it again), the direction of rotation is changed. But it would be inconvenient to have to adjust this manually every time a change of direction is called for. Fortunately, there is a simple refinement to do this automatically. The eccentric is allowed to rotate freely on the axle, but there is a small pin sticking out from it. Next to it, on the axle, and coloured red in the diagram, there is another disc of metal that has half of its diameter taken away, for half its thickness. When the engine is pushed slightly forwards, the eccentric slips backwards until the pin catches against the side of this recess and that allows the timing to make the engine go forwards. After a short run, the throttle is closed. The engine stops and then it is pushed backwards slightly. The pin now reverses against the other side of the recess and the timing is such that the engine runs backwards. This is called slip eccentric gear and is wonderfully simple and rugged. We shall encounter it often.

FIRST PRINCIPLES *continued*

Notice how the bosses on the wheels project slighty beyond the thickness of the rims. This keeps the coupling rods well clear of the wheels. You will see that the crankpin on one side is in line with the centre of the axles, whereas on the other side it is obviously out at the '3 o'clock' position. This is our first intimation of 'quartering' – where the position of the pins on one side is at 90 degrees to those on the other.

You will note that the mainframes go into slots cut into the top of the steel angle that forms the front buffer beam. There are proper ways to cut these slots but, when no one is looking, put two or three blades into a hacksaw – side by side. Use some cutting oil, don't rush the job and it is as quick and easy as anything. Modifying a hacksaw, so that the holding pins are longer, is a worthwhile little job. It will repay the effort many times over in years to come.

Where a round rod has to slide easily in and out of a chamber containing steam pressure (like a cylinder or steam chest) without leaking, we resort to a useful little thing called a gland. Typically, there is a brass hex nut which has a recess inside, into which goes an O-ring. As the nut is tightened it compresses this ring, making it fatter. It then presses against the sliding rod – eventually making it steam-tight, but allowing the rod to move smoothly, without undue friction.

With steam, you can't use any old rubber ring. It needs to be made of a material that can stand the heat. So buy them from a model engineering supplier – not your local garage. Before the coming of O-rings, graphited yarn was used. It looked like silvery black string and made your hands filthy. It was wound round the rod neatly and then screwed in place. Incidentally, if you saw a full-size steam engine oozing jets of steam from where the rods emerged from the cylinders, you could tell that the glands wanted repacking. 'Graph yarn' is still quite acceptable for packing glands, but modellers enjoy the delight of instantly being able to slip an O-ring onto a rod (*see* page 24).

are rugged. So we may well be happy with slip eccentric gear. I once invented a valve gear that allowed slip eccentrics to work expansively. It worked well enough, but was slightly more complex than Walschaerts gear, so was a hollow triumph. A halfway house is a simplified Walschaerts gear that helps unassisted starting via radio control. It is particularly favoured by Roundhouse Engineering for sound practical reasons.

I am going to suggest that the newcomer to the hobby sticks to slip eccentric, or, in the case of 16mm scale, uses ready-made valve-gear. Such things have a lot of joints and pins. Fits of these have got to be very accurate. There must be little or no sloppiness at all, because this will affect valve events.

TENDERS

Tenders will not be touched on in great detail, for the simple reason that, by the time you have built an engine, you will need no new skills to make a tender for it. In the full-size world, they have either inside or outside frames. Because the Southern Railway had no water troughs, they sometimes used particularly long tenders that had eight wheels or even featured bogies. If your chosen prototype has a tender, I suggest that you keep things simple for a first model. Don't worry about suspension or compensation. If you want to semi-permanently couple an engine to a tender, drill simple holes in the drag beams of both and run a long nut and bolt through. Using washers, slip a spring either side of the tender drag beam, so that there is slight springing in tension and compression.

SPRINGING

Full-size engines employ springing on all axles. These run through solid blocks of a slippery metal such as bronze or gunmetal. These blocks are free to slide up and down in guides, often in the form of U-shaped castings called horns. The blocks are fixed to springs, which may either be coil springs or in the form of leaf springs. Leaf springs are often unreliable and unpredictable in small models, so it is common practice to model such springs where visible (such as outside the frames of a tender), but to use concealed coil springs to do the job.

For smoother running on less than perfect track,

Case Notes:

Vulcanette

There has been a whole genre of dainty little 2-4-0 tank engines that have oozed elegance from every pore. In particular, Sharp Stewart and Beyer Peacock were responsible for some particularly graceful examples. This freelance design pulls together all of the elegant features that I like. But, of course, you can easily amend the outline to a specific prototype. This engine is based on Vulcan practice. Around 1870, a particularly svelte example was exported to become Japan Number 1. It survives in a museum to this day, albeit with later modifications, which, in my view, rob it of some of its graceful lines.

So let us consider some of the practical points together. The proposed scale was 1:22.5 to run on Gauge 1 track. You can see from the square end to the frames at the right that it was originally conceived as a spirit-fired model. Those long side tanks would conceal inner wrappers that would provide a substantial heating area. They would also provide a good place for a big chunky whistle. The advent of ready-made gas-fired boilers would seem to make for an easier life, but an off-the-shelf boiler would not be long enough for this purpose. So the actual boiler would be set inside a longer copper tube, with an empty space forward of the working boiler. Going to gas firing would also mean that the rear profile of the frames could be made more to the airier outline of the prototype.

The cylinders would be best scratch-built, with the steam chests inside the frames, through slots cut out for the purpose. If you like this design but are not happy to tackle scratch-building cylinders at this stage, there are a couple of alternatives. The first would be to use a pair of modified Roundhouse cylinders, tilted on their sides so that their steam chests are inside the frames. They would be slightly too chunky in appearance compared to the originals, but some of that could be hidden behind those deep valances (footplate reinforcing strips that also add structural strength). There would just be room to fit one or two working cylinders inside the frames, between the front axle and the boiler, driving the real axle. In that case, the outside cylinders would be dummies.

The cab roof conveniently sits on the side tanks and the rear bunker – and so is an obvious candidate to be made 'lift-offable'. The coal bunker is a natural place to hide a gas tank.

It would be worth putting a bit of side control on that front pony truck, rather than letting it just hang there. The horizontal pivot point can be set back to near the front axle. Soft springs could 'tie' the swivelling arm (not shown in the drawing for clarity) to the inside of the frames; the pony wheel would want to tug the model gently around curves. With those large wheels, it would tend to be skittish – to dash off at high speed. This is a good time to mention a useful way of

> **Vulcanette** *continued*
>
> controlling this more finely. Buy a simple model-engineering steam valve and fit it to the exhaust pipe coming from the cylinders. Cunningly, things can be arranged so that it projects through the smokebox door and thus the door handle is what rotates to regulate the exhaust. Combined with a normal regulator at the backhead, this gives beautiful slow speed running, both up and downhill. It does so at the expense of a nice exhaust sound, but it cures an otherwise over-excitable engine.
>
> As engines like this were produced to mainline standard gauge, a model would look very pretty in 2.5in gauge. In Gauge 1, the small diameter of the boiler might mean a limited duration from a boiler filling, but a narrow gauge version running on 45mm (or 32mm) gauge track would be an attractive proposition.

there is a further refinement. Compensation is an arrangement of rocker arms so that, when there is movement in an axle box due to track irregularity, it moves another axle box to help compensate and to make the riding smoother. If you have ever had the pleasure of watching an old American-built steam engine swaying along some dreadful old track, it is a joy to watch in motion.

But none of this scales down to our tiny sizes very well. It is another of those cases of not being able to scale nature. We don't have the bulk and weight to work for us. Having learnt what springing is and how it works, perhaps it is something best left off your first engine.

SMALLER DETAILS

Handrails

There is a variety of small details in a model steam engine that are simple in themselves but which may be new to the reader. For example, there are handrails fitted in various places, depending on the type of engine. There will usually be vertical handrails either side of cab openings to help the crew climb aboard. In our models they can be bent pieces of steel wire, threaded at each end. You will find that I am partial to suggesting jigs for making things. A little drilling jig can consist of a small piece of angle with two holes drilled in it. This ensures that all four holes are equally spaced, relative to the edge of the cab opening. An alternative is to use bent brass wire and soft solder it to the cab side, using a temporary spacer, to make sure that the handrail is also parallel with the cab side.

The other, more common, type of handrail uses separate handrail knobs. They would be fiddly things to make but are cheap to buy ready-made. For handrails themselves, straight steel wire looks best (as per the prototype). Depending on the scale you are working in and the size of rail required, check out model aeroplane control wire and unwound guitar strings as possible sources. On balance, it is not a good idea to drill holes in a boiler to take handrail knobs. Each hole is a source of leakage and the fitting has to be soundly silver soldered into place. If you are nervous about this, leave these handrails off. A budget tip is to use small split pins as handrail knobs, slipped over a very small washer. It looks perfectly acceptable out in the garden. That elderly LMS engine in the Introduction uses split pins – and I doubt that you noticed it.

Smokebox Doors

To keep the prototype airtight, it uses a central locking device. This consists of two handles that look like the hands of a clock. One works a screw tightening device and the other locks a type of pin (called a dart) in place. The main thing to remember is that the inside handle should always be hanging straight downwards, in the locked position. The other can be at any angle.

An alternative with some engines is a round handle in the centre of the door. A notable user was the London and North Western Railway (LNWR). This was prone to jamming, with all the ash that swirled about in a smokebox.

Overseas locomotives frequently use lots of smaller clamps (dogs) around the circumference of the door. Primitive engines might have small

Front end of a rather elderly model. Note the conventional smokebox door arrangement and the prominent sandboxes – to allow dry sand to trickle onto slippery rails when required.

half-doors that hinged upwards. British locomotives usually had/have door hinges on the right side as you face them.

Headlamps

British standard gauge locomotives, in service, have been required to show headlamps of a particular pattern to indicate the type of train they are hauling – everything from a light engine up to a Royal Train. These lamps may have white or black bodies but will show a white light. The Southern Region of British Railways often used the practice of exhibiting plain white discs in daylight, instead. Many narrow gauge engines merely carried a single headlamp and, if a light engine, a red tail lamp as well.

An ordinary oil headlamp should not be confused with a headlight. These are more akin to the headlight on a car and are meant to light up the way ahead. These were very rare in British standard gauge practice, the most notable exceptions being the Lickey Banker 0-10-0, which had a specific need for one, and some industrial engines. In narrow gauge, there may be overseas locomotives that have been imported, or even repatriated, still retaining their headlights.

It is possible to buy working headlamps that are tough enough to withstand some heat. Be careful where you route the wires to a concealed battery. One little dodge is to use some heatproof insulation sleeving, rescued from an old electric cooker.

Headlamps sit on brackets called lamp irons. In model form these can be folded stubs of flat brass shim (thin brass), soldered in place. If you want a tougher lamp iron, cut a short length off a piece of brass angle. Model lamps tend to be a sloppy fit on irons anyway, with the result that they droop, wobble or both. If they do, glue a sliver of brass shim into the lamp socket. An alternative form of lamp iron – again common with the LNWR – is an actual square socket, which is most easily made from small square-section brass tube.

Whistles

As with suspension, you can't scale a whistle down and get the same results. The note is dependent upon length. Thus model whistles tend to give off a pathetic watery peep. During the history of model engineering there have been designs for resonating boxes, which perform better. If you particularly want to have a whistle in your engine, go for the biggest one that you can hide in the side tank of a tank engine. Because you will be taking unsuperheated steam out of the boiler and through an exposed pipe, the whistle will sound wet and asthmatic at first. Connoisseurs of such things have experimented with keeping the whistle and pipe warm in some way to try to help things. A big whistle sounds better, but may drain the boiler of a model quicker.

Buffers and Couplings

Most standard gauge steam engines will feature a pair of sprung buffers and a centre coupling hook with a three-link chain. That chain may have a

A re-imported narrow gauge locomotive on the Welshpool and Llanfair Railway, hence the unusual headlight and cow-catcher.

screw device to allow the coupling to be tightened up – called a screw link coupling. Primitive locomotives might have had simple wooden blocks instead of sprung buffers; these are called dumb buffers. One refinement of these was to face them with iron plates (in model form, mild steel).

It is a pleasure to watch a model of a standard gauge train gliding around a long oval track, its sprung buffers occasionally kissing. It can be less of a pleasure if you have a single-cylindered narrow gauge engine, hauling a lot of light wagons – especially when shunting. The sprung couplings can build up oscillations along the train until the guard's van is having a terrible time shuttling backwards and forwards erratically. For my narrow gauge interests, over the years, I have come to prefer unsprung couplings

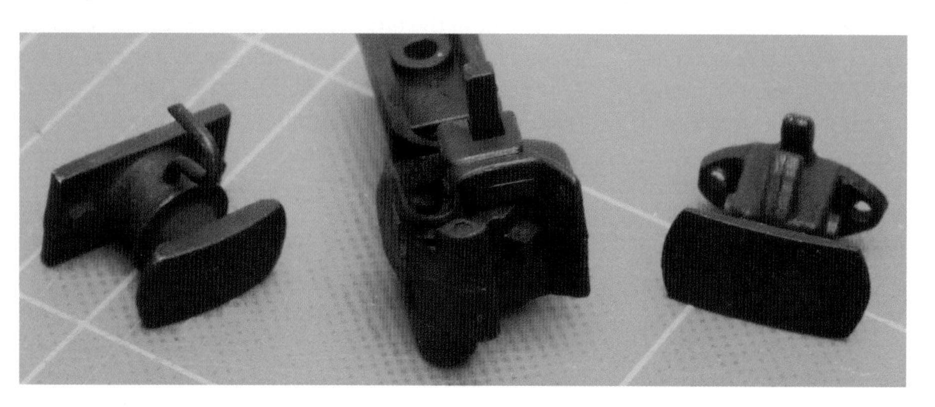

The centre knuckle coupling is flanked by two examples of a centre buffer/coupling.

A whimsical thought by the author for a tiny standard gauge loco as it might have been built by the Great Western Railway.

and buffers. Other people, quite rightly, have formed a different view.

Narrow gauge practice is usually to have a single centre buffer/coupling. These come in a whole variety of patterns, usually based on chopper hooks or knuckles that engage. It is a common convention for your garden railway to standardize on a particular type of coupling, irrespective of what the prototype might have had. After all, that is what a full-size railway would have done. When it bought in second-hand engines or rolling stock, the first job would be to fit them with their own couplings at their coupling height.

Very simple diminutive railway systems, where a few wagons and no passenger vehicles are being trundled around, could use simple hooks and loose chains. There would be a bulge of some sort at each end of the vehicle, with the hook projecting above, to give a clean pushing contact with the next wagon.

Try to get hold of properly sized scale chain, rather than some huge chromed thing that came from a bath. It looks much better. In 16mm scale you can buy the right thing in mild steel, which rusts splendidly. I use a pair of long-nosed surgical forceps for coupling up the links. At several places on my garden railway I keep small reserves of three-link chain. They are wonderfully susceptible to being lost. In days of old, we used old silvered clock chain that we painted in iodine to darken it over time. It is still a useful little dodge to keep stored away.

This concludes our reconnaissance of what things are called and how they look. It will get fleshed out further as the chapters unfold. We will therefore now turn our attention to the basic metalworking that will be needed.

CHAPTER 3

Basic Metalworking

We will look at the basic processes for metalworking with hand tools, in the order they will be needed for building a locomotive. There is nothing difficult; just a few techniques to get used to. If all of this is entirely new to you, then I always suggest having practice attempts on scraps, before you cut into the material you have bought specially to build an engine with. Be patient with yourself: once you get a feel for these jobs, they will be with you for life.

MEASURING AND MARKING OUT

Get comfortable and position yourself in a good light. It is worth having a new metal ruler to work with, as this will be easier to read. (It is better to use a steel ruler rather than a plastic one.) A piece of hardened steel, sharpened to a fine point, is needed for the actual marking. The proper tool to use is a scriber, although I have used a large masonry nail in the past, as well as those old darts mentioned in the Introduction. The millimetres engraved on the ruler look tiny at first, but you will quickly get used to them. Get into the habit of peering down, exactly over the top of them, to avoid slightly misreading where they are. Try scribing a line on the scrap of the sheet metal you are using. It will be thin and accurate – and hard to see. Some marking-out fluid will be required. In years past we mixed shellac and vegetable dye; also, using up old aerosols of paint is an alternative, although this can 'tear' slightly when scribed. A big black permanent marker pen will also work, although it is prone to smudging. Marking blue is still the best option.

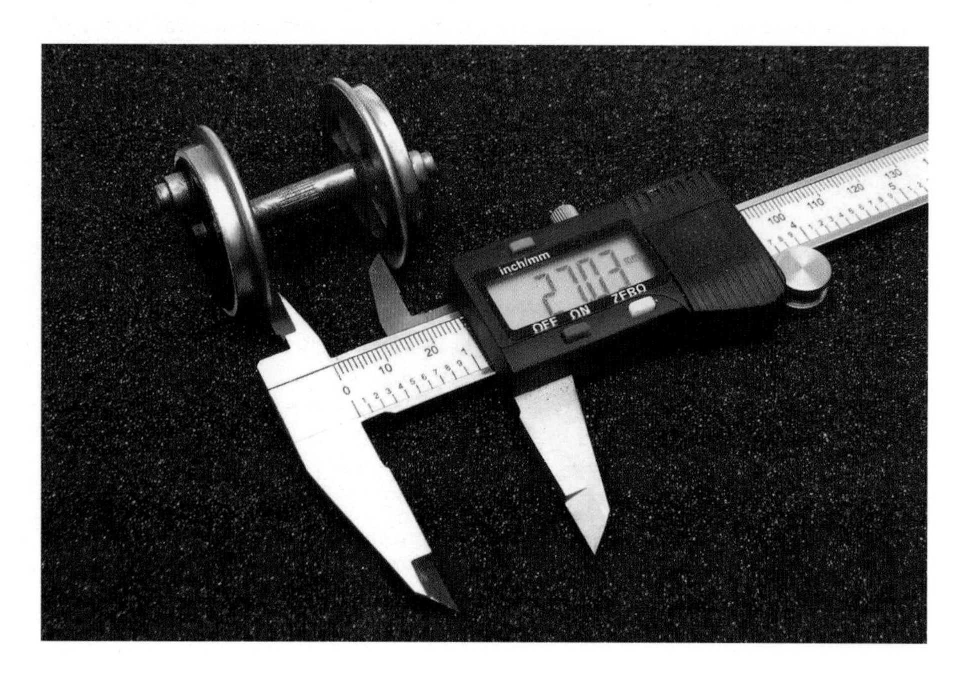

Measuring back-to-back with digital calipers.

Scribing a line along a sprayed paint layer. The adjustable 'fence' was home-made many years ago.

When you lightly scribe a line, it stands out bright and clear. A metal engineers' square is needed to draw lines at true right angles. If it is an old one, found in the back of a drawer somewhere, check that it is still truly square by offering it up both ways along a true straight edge. If in doubt, discard it and buy a new one. There are some economies that just aren't worth making. You will find that a dial caliper will be a good servant to you over the years. One that has a digital read-out is now cheap to buy and makes life much easier. With a square, a steel ruler and a caliper, you can transfer a drawing made up of straight lines onto metal. Curved lines can be marked with a set of dividers. These are the heavier equivalent of those little things we used at school (although I know for a fact that a certain well-known loco-builder still uses the tin ones from his school days).

If you are marking out something that is one of a pair, like mainframes or tank sides, take two pieces of metal, line them up together along straight edges, drill holes in scrap areas and then bolt them together with nuts and bolts. This way, if some non-critical dimension isn't quite right, both items will match.

As you mark out components, very often they will check each other and any mistakes will show up. But don't be afraid to invent some little jig or other to ensure that things *are* right. For example, the top curve on the front and back sheet of a cab is usually a very large radius. Place your metal sheet on part of a flat board. Locate it with a couple of pins that hold the back sheet in place. Pencil-mark a centre line down the metal and then extend that line further across the board. The marking-out device is simple indeed. It consists of a strip of

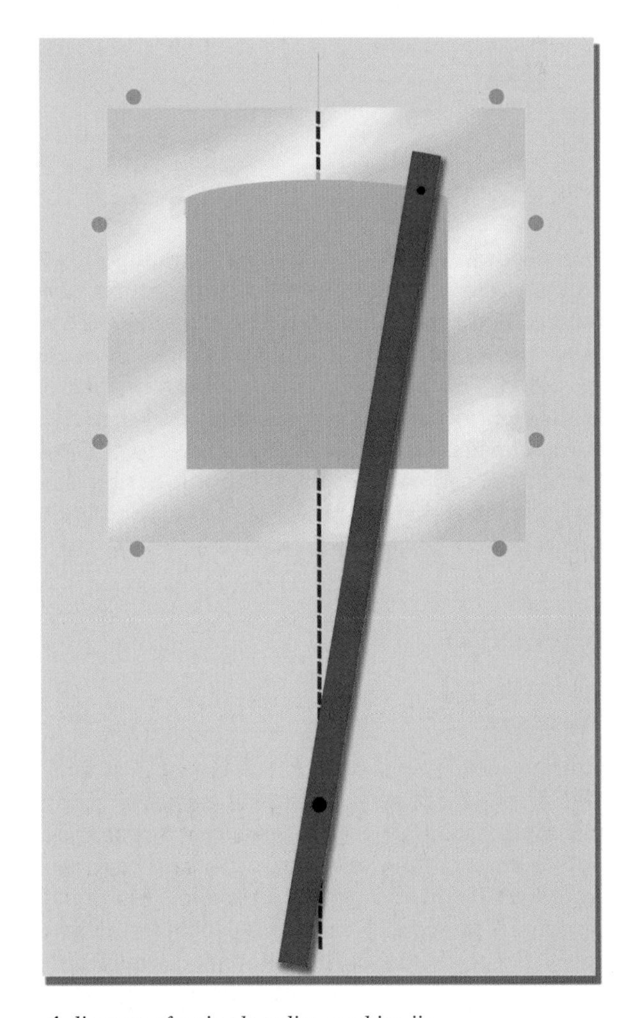

A diagram of a simple radius-marking jig.

CARE OF MARKING-OUT TOOLS

It makes sense to look after your marking and measuring tools. Don't throw them into the toolbox along with everything else. Find a nice flat box with a lid to keep them in. Treat them to a nice piece of cloth to lay on. Give an occasional spray of WD-40 and then lightly wipe it off. In short: treasure them. It is worth it.

Marking a straight line along a boiler.

wood with two holes, spaced apart the radius of the required curve. One hole has a snug fitting nail (acting as a pivot point) pushed through and the other has a similar nail that *just* protrudes through the wood. The 'pivot nail' is tapped into the board at the right distance below the sheet and then the marking nail is scribed in an arc on the sheet. Then change the cab back for the cab front. Scribe this and you know that the radius of the curve of both sheets have to be exactly the same.

SAWING

Hacksaws

Cutting metal seems a very slow business at first. The teeth of a hacksaw blade are each like a tiny planing machine. It is important that not merely one tooth is cutting at a time, which is what you would get if you tried to cut a piece of thin metal upright in the vice. Avoid this by keeping the saw *nearly* flat to the width of the job. You would also use a blade with more and smaller teeth per inch. This will be sold as a 'fine metal-cutting blade'.

Do make sure you buy blades from a proper supplier. The ten for a pound offers in cheap shops are useless, seeming to go blunt cutting ice cream. Use the full length of the blade when cutting – long, slow strokes – rather than hastily stabbing up and down with just the middle of the blade. A good saw blade will last much longer when properly used. Put a drop of oil on the blade to lubricate it, especially when sawing thick steel items. Keep the strokes as level as possible, avoiding a big rocking motion. When a saw blade starts to wear, get a new one. You can cascade the old one down to smaller, less important, jobs. The extreme ends of an old blade will still be sharper than the middle.

Don't use 'flexible' blades for normal work, as the alloy they are made from makes them softer. A proper blade is very tough. However, part of that toughening process makes it brittle at low temperatures and it can merrily shatter when used on a frosty morning.

Much of the foregoing applies to junior hacksaw blades as well. Don't begrudge buying good hacksaw and junior hacksaw frames. These tools will see a lot of work so you might as well be comfortable when using them. Cheap frames often have dubious

Sawing thin sheet, clamped to the edge of a worktop.

arrangements for holding the blade in place, with the result that a blade may want to twist in the frame, which can play havoc with an intended straight line.

They say that the best height to saw at is where a straight forearm is in-line with a straight blade when cutting. I have got used to having the job up higher, but, given a choice, it is better to stick to this wise advice. And let the saw do most of the work; don't press too hard. If it is a long sawing job, set yourself to do whatever number of strokes suits you, then walk away from it. Do something else and then come back to the job. Steam engines for the garden gauges are generally fairly small objects compared to larger, passenger-hauling models. The drudgery factor is therefore much smaller.

I am still unsure about my ability to cut a perfectly straight line every time. This is particularly so when the blade is being held sideways in the frame for a long cut along the edge of a job. I therefore usually cut just a whisker outside the marked line and finish by filing, which also serves to remove the teeth marks. To cut an opening in the middle of a piece of metal sheet, the usual advice is to drill out the corners with holes big enough to accept a blade threaded through and then attached to the frame at both ends. If ever there was a case for the need for three hands per human body, this is it. But somehow the job gets done.

I suggest that you don't drill those holes right at the extreme corners. A slightly misplaced drill and you have little curvy bulges in an otherwise true rectangle. Come inboard a bit. Saw slightly inside the marked lines. When the main cut-out has been removed, finish by filing down to the corners and lines. If you are new to this and a bit worried about accuracy, clamp a sacrificial straight edge of metal against the line and file down onto that.

Fretsaws and Coping Saws

Most sawing jobs tend to be straight lines, but you will be called upon to cut curved lines as well. A hand-held fretsaw, fitted with a metal cutting blade, will do the job but it is not easy at first. The blades are very delicate and break at the least provocation. The ends often slip out of the clamps. It is just a case of being very gentle and patient. There is a distinct 'feel' to using a fretsaw and it takes time to get used to it. This also applies to the other vice that fretsaws have – because the blades are thin, and the frames deep and unwieldy, the cut wants to keep darting from side to side irregularly. I am not an enthusiast for this tool but it will deal with some jobs. A slightly more rugged variant is a coping saw, fitted with a metal-cutting blade.

There is also the jeweller's saw, which can use fretsaw blades but the frame is much less deep; I find this easier to control. Normally, the principle with all blades is that the teeth point away from you so that they cut on the push stroke. But for very

Sawing a straight line against an edge of scrap metal.

delicate work, some people suggest having the teeth pointing backwards so that you don't force a cut too strongly. The pull-back stroke tends to be more gentle.

Life is easier if you have a powered fretsaw, in which case the job is offered up to the moving blade. This provides much more control. Some fretsaws have become very powerful indeed and can cut a delicate pattern through half an inch of mild steel or two inches of wood. The more meaty ones are described as scroll saws.

One tool which seems to have dropped out of favour is the abrafile blade. This looks like a length of flexible wire, which has a roughened, file-like surface all the way round. If held in a frame like a modified coping saw, it will cut in any direction. The blades are flexible but can still snap if abused. I have found these blades to be an excellent resource over the years, and have helped out with innumerable awkward jobs. If you can find abrafile blades, grab a few. You may sometimes find them in the form of tile saws.

Some folk get on well cutting metal with electric jigsaws, fitted with metal cutting blades. I am not one of those. If I do cut thin metal sheet with one, I make a point of backing it with an offcut of plywood – this is good practice for much sheet-cutting anyway. Electric bandsaws can save labour, but I am not going to suggest you buy such things for making small steam engines. A sharp hacksaw blade and a decent file will cope pretty well with most things that a small steam engine throws at you.

Sometimes we can indulge in a little low cunning. Suppose you want to cut something that is a mixture of straight edges and small radius curves. You could drill holes, the radii of which matches the curve you want, and then saw/file straight lines to join them up. Sawing and filing are right at the heart of metalworking. They seem laborious at first, but aren't really. Using long, slow strokes with the hacksaw, letting the blade do the work – and a whiff of cutting oil – and the job is soon done. Filing equally benefits from steadily working into the metal, without too much pressure.

FILING

Files come in all shapes and sizes. However, there is no need to build up a vast collection before you get started. A handful of sharp files will be far more use to you than a drawer full of rusty horrors found in a derelict shed. The basic shapes are flat, round, square, half-round and triangular. They come in a variety of 'roughnesses', but general-purpose files will do for most of our needs. In time, you will come to learn about the different 'cuts' of a file, but there is no need to worry about it right at the start.

There is a whole world of tiny needle and riffler (tiny with slightly curved ends) files to investigate. I have a collection of such things. Occasionally, one or two get used. Again; that same principle: don't expect much for a tuppence-a-dozen pack. Buy a few good ones from a proper supplier. You will notice that such advice is repeated in this book, as well as elsewhere. In fact, if you go back through model engineering literature for a hundred years, you will find this theme reprised. It is as true today as it ever was.

The Art of Filing

You would think that filing would be a straightforward activity: something that you just do. However, there are rights and wrongs to it. Filing a flat, on a job held in the vice, is best achieved by keeping the file as horizontal as possible. If you are right-handed, hold the file in your right hand and steady the far end with your left. Press on the forward stroke and try not to drag the file backwards under pressure. If you are filing a long flat, moving the tool sideways as you push forwards helps to spread the cutting action more evenly. You will probably be filing down to a marked line, so stop to check things frequently.

If you really are worried about your ability to file to a straight line, there are two dodges. The first is to use a sacrificial straight edge – an offcut of thin metal that has a true straight edge. Clamped next to the job in the vice, you can feel its presence. Paint its top edge with black marker pen. By filing diagonally sideways, you can file down to it until that ink is just starting to be affected. Final smoothing is achieved by moving a file sideways across the length of a job.

Filing an axle box to fit an opening. The marking fluid came from a green permanent marker pen.

The other trick is to clamp a large file in the vice and drag the job along it, trying not to rock it, although you can still apply a little extra pressure at one end to correct any unevenness. You should aim to become proficient with a file, but until you reach that point, these dodges may keep you going.

You may notice that some straight files have smooth (safe) edges, whereas others have cutting teeth. You might use a smooth edge if you were filing out an opening where it is important not to make even the slightest cut sideways. If you want to file a curve on a piece of metal sheet, don't try to make the file follow the curve; instead, rock it in the opposite direction. However, the easy way to get an accurate radius is to use a button, which is a short stub of metal of the right radius. Clamp it next to the material to be filed and file down to that radius. Strictly speaking, the button should be able to rotate somehow, so that you don't file that as well.

For your first locomotive, there is a useful dodge to file a curve that is bad practice. I keep a box of old ball races, of all sorts of sizes, collected over the years. The outer curves are made of a very hard material. Clamp one of the right size next to the job and file down to that with a tool that is kept just for this job. It will skate over the very hard metal but smooth down the edge of the workpiece. It doesn't do the file much good but it is only once in a while and, besides, no one is looking. Try to make a point of fitting wooden handles to files. It is kinder (and safer) for you and makes them that bit easier to use.

Internal curves are dealt with by using half-round files. One dodge for sawing and filing *very* thin metal is to sandwich it between a couple of sacrificial pieces of thin plywood, the top one being marked to outline. Needless to say, there are many jobs where you will want a matching pair. You will therefore have two sheets of metal held together by a couple of rivets or screws – or even glued with superglue – to be released with the blowlamp afterwards.

SOLDERING

You may well be familiar with joining a couple of wires together, using a soldering iron and some cored solder (that is, with its own flux built-in). A little 25W or 40W iron does the job nicely. But such things don't usually cope with chunkier pieces of brass or copper. Heat is conducted away faster from the metal than the iron can put it in. For some platework jobs, you can soft-solder thin sheet metal by using a much bigger and more powerful iron. In my dim and distant past, I used a big 'dumb' iron, heated on the gas stove. I still have it and it gets occasional use.

Nowadays we enjoy bottled gas blowlamps that will heat the metal up directly. These are a great improvement on the fierce pump-up paraffin or petrol things. For most of our modest jobs, a burner that screws into the top of a gas tank does the job, but for larger tasks like boilermaking, we use a torch that is connected to a separate bottle via a hose and an approved regulator.

Either way, the strength of these devices can be enhanced by means of a hearth (*see* page 40). This is a metal-lined enclosure on which refractory materials are stacked. These glow red hot and reflect the

A small blowlamp will cope with all soft soldering jobs and some small silver-soldering tasks.

heat back into the job. There are two secrets to successful large soldering tasks. The first one is getting enough heat where it is needed. The second is cleanliness.

Pieces of metal should be scrupulously clean. As well as physically cleaning the metal, a flux is used to prepare the metal chemically. A typical flux for soft soldering would be Baker's Soldering Fluid. This is a nasty substance to handle and it is also corrosive, so soldered components need a thorough scrubbing afterwards to remove every trace of it. Indeed, it is so fiercesome that it is falling out of favour.

Soft Soldering

Let us look at the process of soft soldering. To start with, we should make note of the fact that all soft solders are not the same. Although they are alloys of similar metals (tin and lead, sometimes with a small addition of antimony), the proportions vary. Typical plumbers' solder is 70 per cent lead (the cheaper metal) and 30 per cent tin. Where a stronger soft solder is called for, the proportions go to 60 per cent tin and 40 per cent lead. For even higher quality, antimony (about 6 per cent of the tin content) is added. A solder containing antimony should not be used on brass or zinc, as there is a risk of making the metal brittle.

A variety of soft solders with a high melting point is available and there are specialist solders for use with zinc-based die casting, although you are unlikely to need them for making small steam engines. An ordinary stick of soft solder will cover most of our needs. Mention should be made of solder paint, however. Paint this onto the areas to be joined. Hold the two components closely together and then play the blowlamp over them. It is a quick method and a useful one.

For electrical work (possibly in connection with fitting working lights or with radio control), it is essential to use a non-corrosive flux paste or a non-corrosive cored solder.

The actual act of soft-soldering items together is quite simple. Make sure the brass or copper pieces are really clean, apply a spot of flux (I use a cocktail stick) and 'tin' them. This consists of melting a very thin layer of solder onto both parts. Clamp them together (bulldog clips or metal hair grips are useful) and reapply the heat, feeding in a stick of solder. If the solder just 'blobs' then the metal wasn't clean enough, or there was not enough heat.

On the other hand, it is possible to overheat a job to such an extent that the solder won't take. If this happens, leave it to cool down and scrub the metal clean, before starting again. Soft soldering is more a knack than anything. After a few practice attempts, you start to sense the right amount of heat to apply, from the way the solder seems to liquefy and flow into a nice smooth layer. If you have used a corrosive flux, make sure to scrub the joint clean straight afterwards – try using a kitchen scouring powder and an old toothbrush, or leave the job soaking overnight in water to which a denture cleaning tablet has been added.

You will be surprised at how little solder is actually needed. Indeed, your first attempts may well result in too much being used – and you will be scraping and filing the surplus off. But, as ever with these things, proficiency comes with practice.

Having now read about percentages in alloys and the like, you can forget most of it for now. Most soft soldering is done with 'ordinary' soft solder and a non-corrosive flux. If and when you move on to wider techniques, you now know what to look for.

Hard (or Silver) Soldering

It is generally reckoned that the dividing line between soft and silver soldering is at around 750°C.

Most silver soldering can be done with a stick bought from a DIY store and some borax powder from the chemist's shop. But, having already read of some of the permutations available to soft solder, it will come as no surprise to discover that there is a bit more to the subject than that. The ability to hard solder a joint and then to hard solder another joint nearby, without melting the first, is dependent upon using different silver solders at progressively lower temperatures. This is more applicable to large-scale passenger-hauling engines.

Silver solders, unsurprisingly, contain silver. This means they are expensive, although a stick goes a long way. You *should* be able to silver solder a simple boiler with half a stick. They also contain varying amounts of cadmium, copper and zinc, but you will be relieved to know that I am not going to quote percentages at you again. Silver solders are known more by trade names, such as B6, Easy-flo, Argo-flo. The usual supplier and fount of all knowledge is Messrs Johnson Matthey. You can find details of them on the Internet or in model engineering magazines.

BRAZING

For very tough jobs, brazing is available. This takes the temperature range higher. Brazing strips have melting points between 750°C and 1,083°C. The reason for this specific top figure is that this is the melting point of copper. I am going to suggest that brazing need not be in the beginner's bag of tools for building small-scale locomotives.

For basic jobs, we have already mentioned borax powder. Mix a little of this into a smooth paste, using a drop of water, and spread it on the area required. But for specialist hard solders, we need a specialist flux. Easy-flo is the all-purpose, fluoride-type workhorse. It is good for copper, nickel-silver, brass and – of interest to us – mild steel. Its value is that it is a good solvent for most metallic oxides.

The Hearth

A hearth is needed for silver soldering. This is a partly enclosed area with some heat-reflecting bricks, so that the temperature can really build up around the job. These can be bought ready made, but it is possible to make a small hearth for yourself. Line an area of the workbench with some sheet metal. A useful dodge is to make use of an old cast-iron firegrate. Put a piece of sheet metal against the back and the base. However, the real business is done by refractory bricks that go onto the 'floor' and 'back' and form some sides: in other words, a basic armchair shape. Using a hearth, you will get much more heat from any given flame.

I also keep a few irregular small pieces of a firebrick close by. When you are silver soldering a small component, you can pack these pieces around the job to reflect the heat back into it. These bricks can sometimes be quite soft and it is easy to drill a hole into one. This is useful for holding a bit of rod steady. No *not* use pieces of breeze block or stone, as these will explode under fierce heat. In my youth, I used firebricks from an old domestic fireplace, but a few proper refractory bricks do not strain the budget.

It is important to make sure that jobs are held together quite firmly. As well as there being every chance that you could nudge something during the operation, there is also a rather blustery jet of gas coming out of the torch. Old hairclips and metal stationery clips can be used. A good standby is to bind parts securely together with thick steel wire (too thin and it can burn through).

The gap between parts, for hard soldering, is very small: typically 0.001in to 0.003in, although there is no need to be precise, in practice. There is a good capillary action with silver solder. One important requirement is for the heating of the metals to be as quick as possible, so as to avoid liquation, which is when slow heating causes one metal of a soldering alloy to melt first and run away, leading to problems. So, quickly in with the heat, do the job and get out again.

SAFE SILVER SOLDERING

Always do silver soldering with good ventilation. Fluxes can give off rather naughty fumes, particularly if overheated. Silver solders, containing cadmium in particular, absolutely demand that you have good ventilation.

Let us conclude this section by silver soldering an imaginary joint. Two pieces of metal will have been coated with flux, in the areas where the joining is to take place. Some people say that making a pencil mark may stop the flux, and hence the silver solder, from creeping beyond the location required. Both pieces can be held together with a clip or two. The job is put on the hearth and some refractory material placed immediately behind and on either side. With the torch in the right hand (for right-handed people) and the silver solder held in the left, splash the vigorous heat onto the job. If it is steel, get it so that it is just starting to glow red.

Apply the tip of the silver solder to where it is needed. If the job isn't hot enough, the tip will bend slightly and maybe a blob will fall off. What *should* happen is that the silver solder will immediately 'flash', instantly flooding out in a very thin layer. If all has gone well, that's all that is needed. Leave it to cool. If it is not a critical job strengthwise, you can pick it up with a pair of tongs and quench it in cold water. We have already spoken about cleaning all the residues off. If you have been making a copper boiler, this can be cleaned up by dumping it in a pickle solution overnight. (You can buy the proper solution, but I will admit occasionally to using a jar of vinegar recycled from pickled onion duties.) Finally, rinse the job in plenty of cold water.

HOLDING

It may seem odd to devote space to how to hold work pieces – including nuts and bolts – but if you are new to small-scale metalworking, it can seem quite difficult simply to manipulate tiny components in your fingers. Moreover, it is something that rarely gets written about. It is not that the subject itself is difficult; it is more a case of it being new to you. There are the obvious aids like toolmakers' clamps and pin vices, but they don't help you to pick up a tiny nut and fit it on a screw that you might not be able to see the end of (lick the end of your finger and press it hard down on the nut; this will then stick long enough for you to lift it up and get a thumb under it).

Here, a cocktail stick with a spot of grease on the end comes in useful, although there are several other unofficial tools that are valuable. The first little three-wire gripping tool I had was branded a 'part picker' and I still tend to use that name. Small surgical tools like clamps, Spencer Wells forceps and the like are extremely valuable for awkward assembly jobs. You will find yourself faced with many assembly tasks that are entirely new to you: I still keep finding new ones after fifty years.

A selection of clamping tools.

The humble G-clamp and small mole grips are extremely useful.

It may involve holding a tiny screw in a clamp, held upright in a vice and then threading the job over it with the left hand – and then starting a nut off with a greased cocktail stick with the other hand. There is nearly always a way of doing a job. It is just a question of working out how. And once you have done it, mentally file it away for future reference.

Probably the thing that will seem the most difficult is seeing tiny pieces, together with nuts and bolts. Take a look at where the wings of a pair of glasses join onto the front. You will see what may seem impossibly tiny screws holding them on. But, I promise, you will soon get used to it. On the subject of glasses, the best close-up aid is a pair of reading glasses in good condition (the plastic lenses in common use today scratch easily, so cosset them). I found that a good pair of reading specs are far more use than magnifying glasses, headband magnifiers and the like. This is particularly true if you work on a light, plain surface that is well lit. This may all seem quite trivial compared to learning complex lathe techniques, but, believe me, getting the business of seeing and holding right makes all the difference.

DRILLING

This apparently simple procedure has its rights and wrongs too. Drilling thick metal by hand is doubtless good for the soul but is also laborious. So I am going to invite you to buy a bench drill. Fortunately, a basic five-speed one is very cheap and is far more use than a hand-held electric drill. For a start, it guarantees that, with careful setting up, the hole you drill will be at a true right angle to the workpiece. A vertical drill stand for a hand-held electric drill is better than nothing, but is usually

A nice little metalworking project is to make small G-clamps out of large nuts.

The part picker: one of the most useful tools of all.

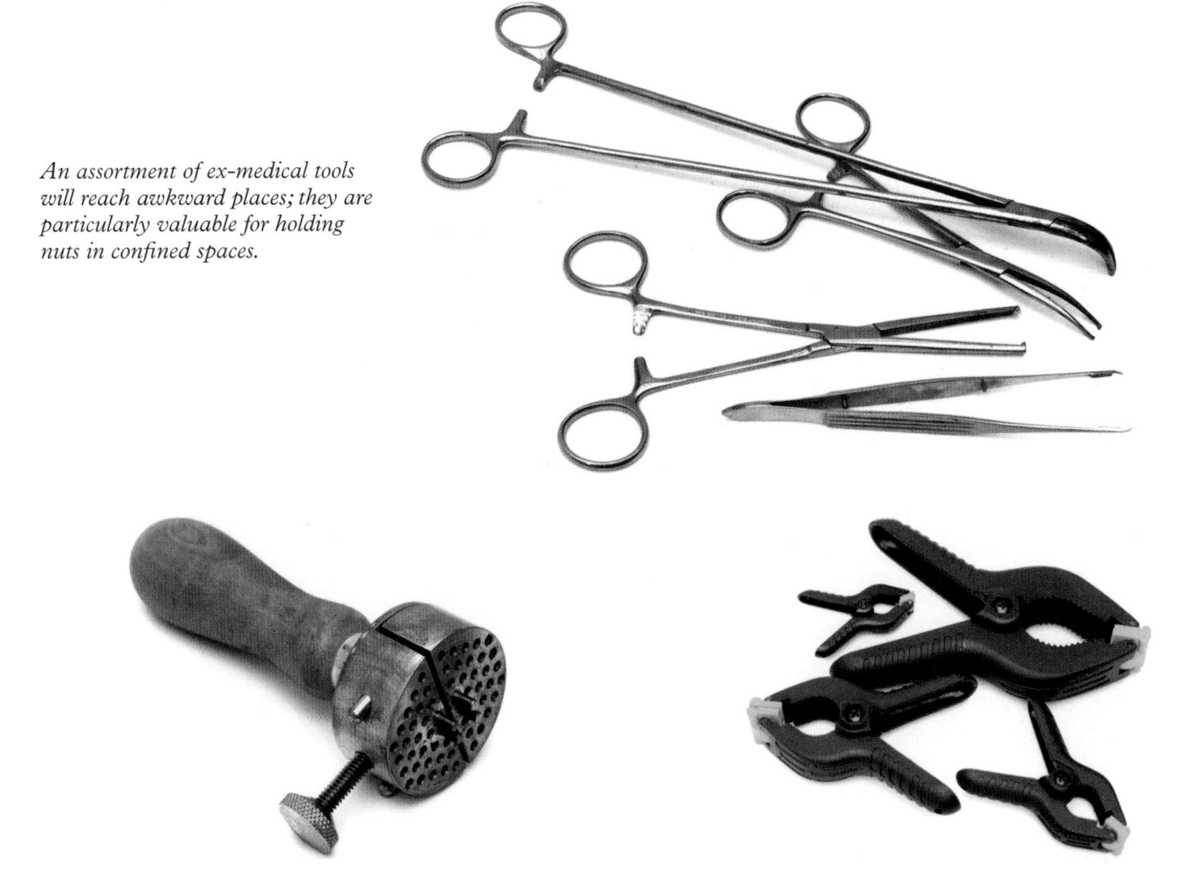

An assortment of ex-medical tools will reach awkward places; they are particularly valuable for holding nuts in confined spaces.

Hobbies suppliers can provide a part holder. It grips odd shapes by pushing pins into holes as required.

Cheap clamps like these are handy – as well as bulldog clips and clothes pegs.

Case Notes:
Java Steam Tram

Whilst basic metalworking techniques are being examined, there is already enough information to hand to build a rudimentary little steam engine. I built the Java steam tram many years ago. An examination of the drawing reveals a simple little Mamod stationary engine at its heart. There is a two-stage reduction gear down to a single axle. I made the mistake of using pulley wheels and Mamod-like spring belts. This worked very well, but the slightest lick of flame would destroy the temper of the belts and they fell slack. What I should have done was use Meccano sprockets and chain, and this is what you see in the drawing.

There was a prototype for this design. It was found as a rusting hulk, with jungle growing up through it. One theory is that it was originally a steam crane that was converted into a 'locomotive' by servicemen during World War II for temporary use.

The chassis of the original model was made from bits of angle, bolted together. The body was thin aluminium. There is nothing significant in this – I made it with what I had to hand. The little platform at one end slides out to allow a flap to be lifted for a burner to be inserted. This was a simple brass box, folded up and soldered. Inside was some fireproof wadding, held in place with a mesh. An old Mamod spirit burner will fit in place, with the handle cut off, although it needs handling with long-nosed pliers. Safety regulations have outlawed spirit-fired toy steam engines. Instead, we have had ghastly fuel pellets inflicted upon us. Despise and ignore them. Because we are building models, not toys (though my wife will debate that actively), always replace them with spirit firing.

There is an opening on top to allow very thick steam oil – a green treacle-like substance – to be blobbed on to the joint between the cylinder and the fixed block. This is a useful way of oiling simple oscillating engines. The thickness means that it will lubricate the moving face and the piston for around ten minutes. I made a bent tin 'catcher' to direct the exhaust steam so that most of it would emerge from the chimney.

Java Steam Tram *continued*

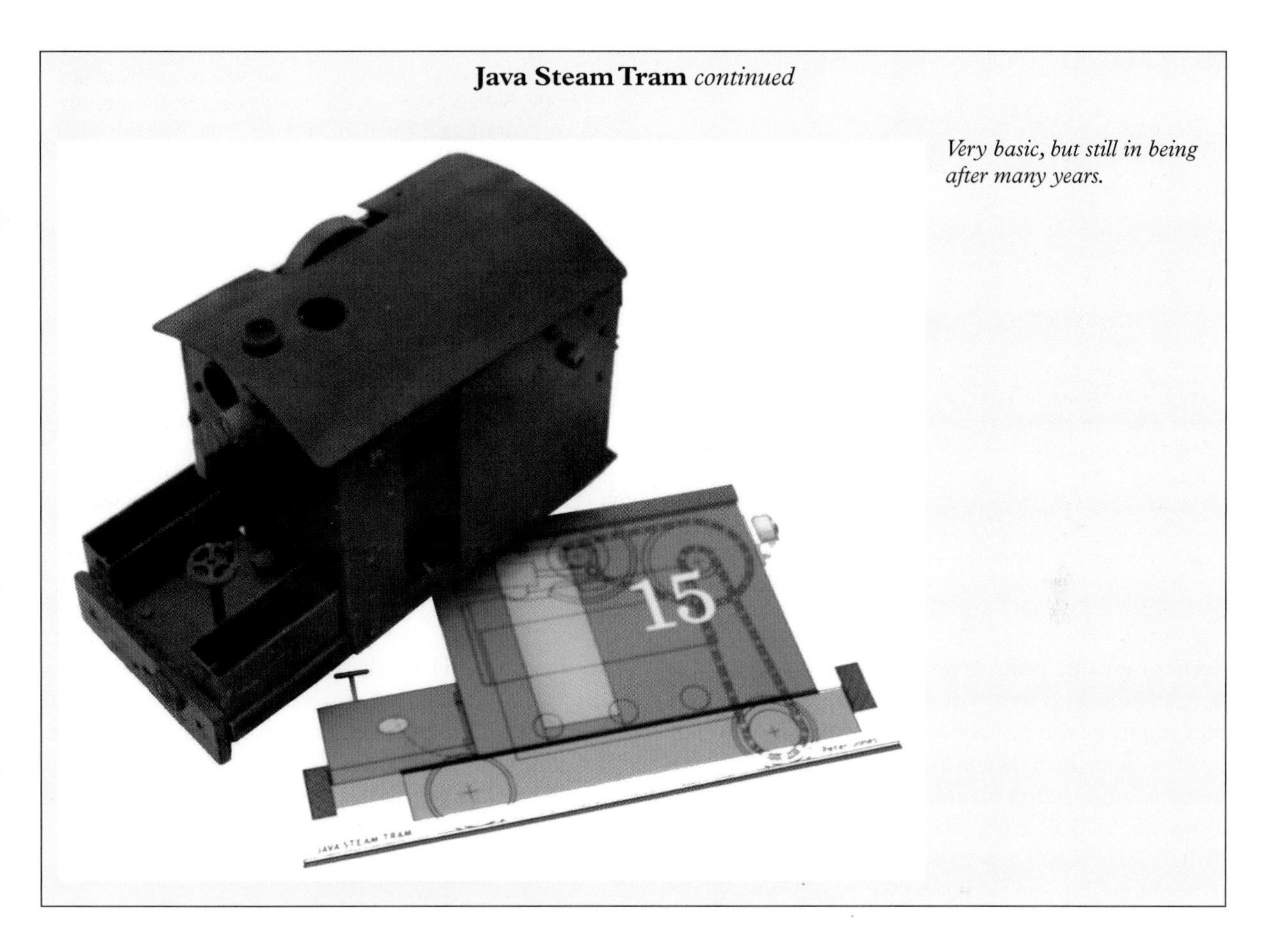

Very basic, but still in being after many years.

much less accurate. The process of 'careful setting up' usually consists of holding a piece of metal rod in the chuck, then putting a metal square on the tilting table and making sure that the vertical arm lines up with the rod.

Those five or more speeds are important too. A drill bit has an optimum cutting speed for a particular metal. A tiny drill whizzes round quickly, frequently withdrawing to avoid jammed flutes breaking it. The outer part of the cutting edge of a large drill will describe a much larger diameter circle and so would cut metal very much faster. So the belt is adjusted to run it at the slowest speed. I could give you pages of tables of cutting speeds in various metals but, for most practical purposes in our simple work, you soon develop an intuition for what feels right or wrong. Until you do, it is better to err on the side of going too slow than too fast. Trying to cut too fast, more often than not, just wears away the tip of the drill. It feels hot and 'wrong'.

Miniature electric drills – like a Dremel – are nice things to have for all sorts of jobs, but, even with a vertical drill stand, they are too dainty for really meaty tasks. For good general-purpose drilling, a cheap bench drill will take care of all your needs. While an expensive deluxe one would be all very well, it isn't essential for our modest endeavours, despite what an advanced model engineer might – with all good motives – suggest.

In an efficient world, all drills would be metric, but the real world has a past. The principal range for model engineering used to be the letter/number series and many locomotive designs feature these. I grew up with them, as did many others. You will find a table of metric equivalents in the Appendices, but you may feel it worthwhile to become familiar with this range.

REAMING

From left to right: an end-mill, a Slocombe drill, a parallel reamer and a taper reamer.

Sometimes our little steam engines will call for a perfect hole with ultra-smooth sides. Typical examples might be the bore of a cylinder or holes in axle boxes where the axle runs through. The holes need to be reamed, after drilling or boring. A reamer looks like a drill with straight flutes. The hole is drilled fractionally undersize and then the reamer is applied. This can be done by hand, with a tap hole or similar to turn it, or it can be applied via the bench drill or tailstock chuck in the lathe. With cast iron or brass/gunmetal it can be rotated at a slow machine speed. The trick is to set it in motion and then feed it into the drilled hole until it has gone right through. Don't stop it then, but keep revolving as you pull it out again. It is all one continuous movement. That hole is now absolutely perfect; it can't be anything else.

The big mistake is to regard the reamer as a drill designed to open out a hole further. This is incorrect – its sole function is just to smooth out any fine scratches caused by the original drilling and to ensure that the diameter is exactly what it should be.

SHARPENING

It is possible to make a single, simple steam locomotive without knowing how to sharpen things such as drills. With only a few holes to drill, a tool may stay sharp long enough to see you through. But there soon comes a point when the cost of a small bench grinder becomes a very good investment. Grinding properly is not a mysterious black art, but it has skills that have to be learnt. The subject will be covered in Chapter 6.

CLEANING-UP METAL

Emery paper is tough. It will scratch many hard metals, so err on the side of using very fine-grained material. Wet & dry papers act like their name. They are slightly dampened and form a fine grinding paste when being used for rubbing. The very finest grades are virtually a flour consistency.

Major surface gunge can be removed easily by holding wire cups and discs in the bench drill and offering the job up against them. This is much easier than trying to hold the weight of an electric drill in your hand against an object held in the vice. But be aware that this will cause scratching, which may have to be cleaned up. I say 'may' because I believe that metals, for working steam engines, which are going to be painted, should not be burnished to a mirror-like perfection first. If you do have such a smooth finish, go over it with the finest emery paper to create a surface of microscopic scratches. This will help the paint to bond well over many years. It

Case Notes:

Bunty

There has been a large number of small, vertical-boilered engines built in both full size and model form. In particular, we tend to use the phrase 'De Winton' in much the same way as 'Hoover' is a synonym for vacuum cleaner. This little North Wales company produced some pretty little engines that were so small you wanted to take one home as a pet.

To give some food for thought, whilst we are looking at simple metalwork, here we have a sketch of Bunty. The original used a cylinder assembly and a reversing block off a Mamod steam engine, but the permutations are endless. There is a reduction gear to drive an axle. Small oscillating engines are notorious for running at high speed with low torque. There have been some examples of these small engines that have had multi-stage gear reduction, but unless perfectly set up there could be problematical losses due to friction.

Another possible obstacle is in getting enough steam generated. A spirit flame needs height to reach its full heat, so the bottom portion of the apparent boiler may just be a sort of firebox. The central flue tube needs to be fairly large. One dodge is to make a spiral of flat copper strip that drops down into the flue. This makes the hot gas swirl as it rises and so it dwells longer. The copper strip also gets hot and acts as a radiant heater. Given a choice, I would be inclined to have a safety valve that is set at 25lb per square inch at least. The low pressure of a Mamod safety valve usually means fairly puny running.

If the boiler is silver soldered, it will stand up to the extra heat that burning methanol instead of traditional methylated spirits provides. It is worth lagging the boiler with some hardwood strips, retained by two or three boiler bands. Coming absolutely up to date, I have found that rubbery silicone baking mats (seemingly usually blue) withstand high temperatures well and make good boiler insulation, with thin metal lagging outside. Keep pipe runs as short and as neat as possible. The traditional approach to a spirit burner is to fit a tank at one end (hinted at on the right side of the diagram) and then to run a feed pipe to several wick tubes in the firebox area. A simpler approach would be to have a small circular 'tin' (tobacco or toothpaste tins were used in the past) that had some wadding in, with a bit of mesh on top to retain it.

This loco will get hot and will probably only have a duration of ten to twelve minutes' running time. But it

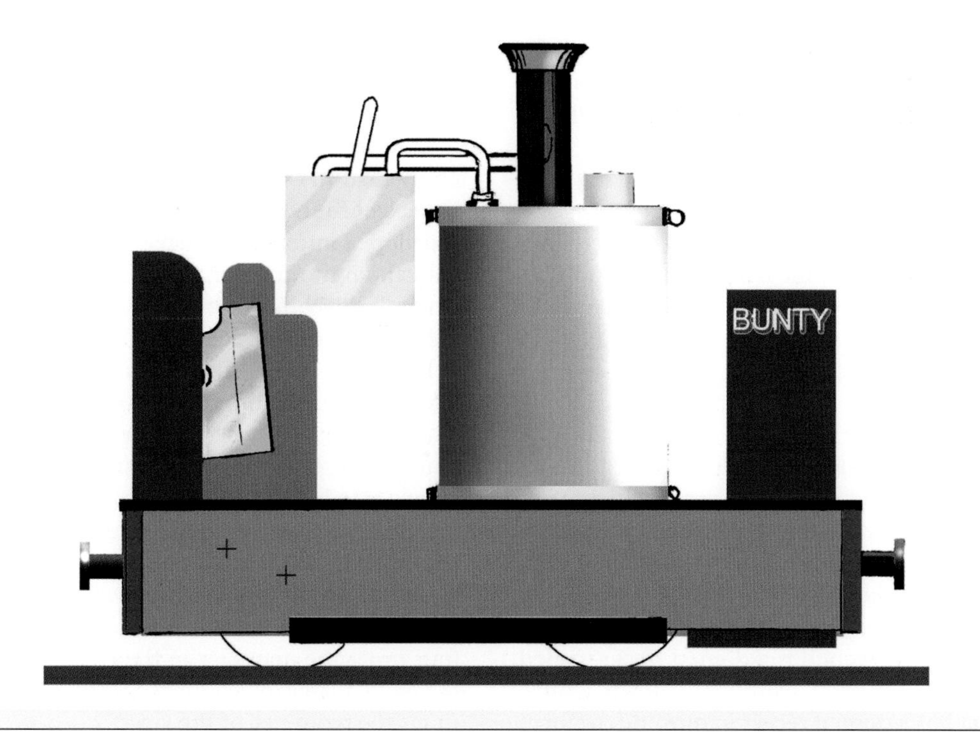

Bunty *continued*

is easy to build and can look quite dainty trundling round with a few slate wagons in tow. There are many variants on this theme – some have fatter boilers, which give better heat generation and a longer run, but which look less elegant.

In the USA there is a whole family of variants that feature a small commercial model boat boiler and engine. Known as BAGRS after their birthplace, the Bay Area Garden Railway Society, they seem to be continually evolving. My example had the problem that the original boiler needed a high chimney stack to draw up the gases. This was fine for a model boat, but if you cut it down to go under bridges it would choke itself off. It also featured a 'safety valve', which consisted of some heat-resistant flexible tube that would blow off the stub of steam pipe if the pressure got too high. However, they can be made to work reasonably well, so do not be discouraged by my own lack of enthusiasm for them.

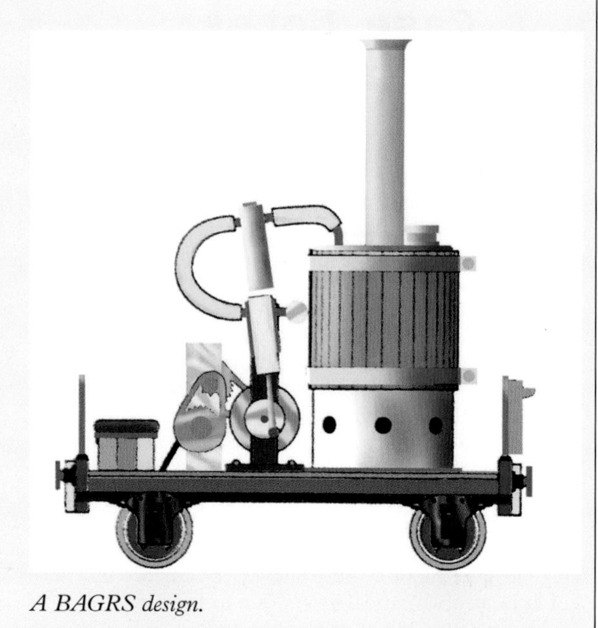

A BAGRS design.

is the same principle as applying an etch coat to brass to help the subsequent paint to adhere.

But where a high polish is called for, conventional metal polishes can be rubbed on by hand. I like to put a big disc mop in the bench drill and use various cutting agents. Fine automobile valve grinding paste cuts well. There are the proper products like jewellers' rouge, but toothpaste is a good substitute for very fine work. Whatever material you use, protect the stem of the bench drill from flying paste. It is meant to be abrasive, after all.

RIVETING

Riveting is a common way to hold parts of full-size engines together and it has its uses in model form. For our simpler needs, there will be two sorts of rivet. The first is a countersunk rivet, which calls for a hole to be countersunk and then the projecting part of the rivet hammered flat until it fills this recess neatly. The second is a domed rivet – called a snap head – which is hammered down with a punch that has a domed recess in one end, to form the nicely rounded pimple we are all familiar with. That hammering down has to be done against an upright 'dolly' consisting of a piece of steel held vertically in the vice, which also has a domed recess.

Dollies and punches are cheap to buy or are easily made. A shallow hole is drilled in the end of a piece of round steel and then a small ball-bearing is dropped into this. The steel is gripped in the vice, with the bottom resting on the vice cross-piece. A single hard hammer blow should turn that sharp angled hole into a recessed dome shape.

Rivets for model use will usually be copper or iron (although aluminium ones have their uses). They may typically be 1⁄16in or 1⁄8in, or their metric equivalents (1.6mm or 3.2mm). They should fit snugly into holes drilled for them. For spacing and location, you may well refer to the drawings of your locomotive. But in the absence of these, there are a couple of simple rules to follow. The space between rivets should be at least twice, but not more than four times the diameter of the rivet. The overlap of the metal plates to be joined should be at least three times the diameter of the rivet.

A selection of rivets and punches.

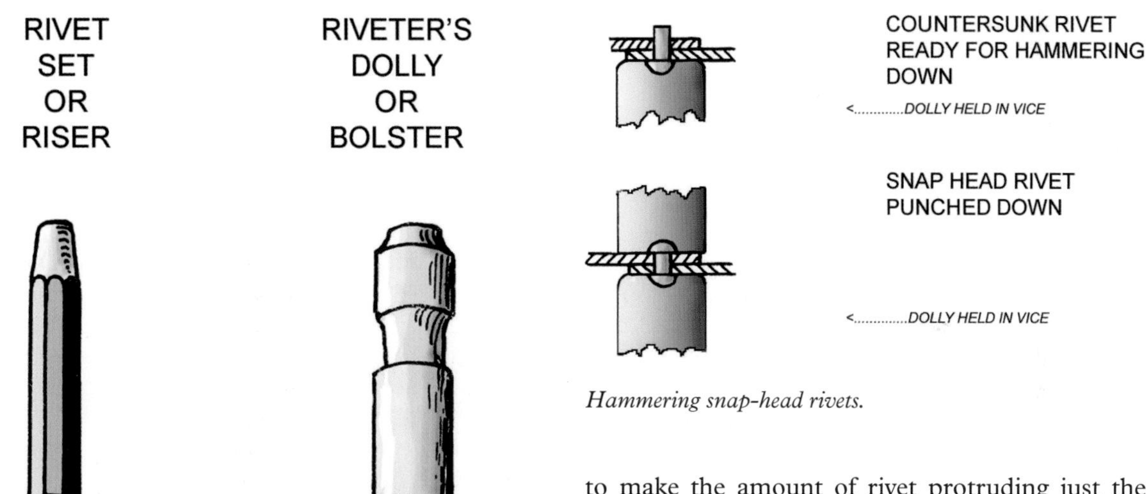

Sketch of a punch, compared to a larger vice-held 'dolly'

Hammering snap-head rivets.

Rivets should be in straight lines. A wobbly row of rivets is an offence against humanity. A simple jig can be made very accurately by using a piece of Veroboard – the sort of thing with thousands of tiny holes drilled in for electronic circuits. Put a spot of white paint on the holes required and then pilot-drill through with a small drill. The result is lots of straight lines and right angles where required. The pilot holes in the job can then be drilled out to the right diameter for the rivet.

Model rivets don't need heating up, in the fashion of old films about shipbuilding. It is important to make the amount of rivet protruding just the right length to fill the recessed cup, or, in the case of countersunk rivets, the amount of the countersink already drilled.

One quick way of getting this roughly right with small rivets is to cut them off, in situ, with a big pair of end-nipping pliers. The bevel on the jaws of these leaves a certain amount of material standing proud. If fortune is with you, this may just be the perfect amount of metal required.

CUTTING THREADS BY HAND

If you buy a set of taps and dies from an ordinary high street shop, it will probably feature metric threads for general purpose use around the home.

Case Notes:

Sussex

Now for something substantially different. Aveling & Porter, in particular, produced rail versions of traditional road traction engines. I have faint memories of being taken somewhere and seeing one of these behemoths lumbering out of the mist and murk on a dismal day. I later went on to hack youthfully at a Mamod traction engine to produce a 4in gauge engine – that gauge because of the overscale width of Mamod fireboxes and bunkers. It worked quite well, too.

Sussex is a design that tries to capture the ponderous nature of that childhood vision. It is freelance but easily capable of being adapted. You will see that a single slide valve cylinder drives a shaft fitted with a flywheel. Movement is transmitted to the driving wheels by means of sprockets and chain. Meccano would be up to the job but my leaning would be towards motorcycle timing chains. If I were to build another Sussex, I would be inclined to fit a second shaft atop the boiler, to allow for additional gear reduction. There is also a faint pleasure in having a flywheel that goes round in the opposite direction to the wheels!

If one of these additional gears could be slid sideways along its shaft, then the flywheel could go round on its own. Many years ago, in the early days of the Bluebell Railway, I saw the rail traction engine 'Blue Circle' sitting in a siding after a day's run. The flywheel was being left to slow down of its own accord, a process that took some hours.

You will see evidence of spirit firing, but gas could equally be used. Indeed, the easy-going nature of a design like this allows for all sorts of levels of engineering and technology. Do not regard this design as set in stone. I once build something like this in 5in gauge, out of scraps. Ah, the follies of youth; how sweet they seem now.

Building a small steam engine will call for something more precise. There is a range of small threads that are called British Association (BA). The largest is size 0 and they get finer and finer as the numbers increase up to about 12BA or 14BA. The last is so tiny that you may need a magnifying glass to see it clearly. Typically, model engines of our sizes will use 2BA, 4BA, 6BA and 8BA quite frequently. An interesting sidelight to model engineering is that the gloriously traditional Stuart Turner stationary engines often used 5BA and 7BA to cut threads in holes – and then fitted screws

A selection of taps in a tap holder.

whose heads were one size too small. This presents a particularly delicate appearance.

There are some applications where a very fine thread is needed, with a lot of turns per inch. An example would be where the thread is part of a screw-in needle valve and fine control is called for. Here, the model engineer (ME) thread series is used. There is no need to buy full sets. Buy the individual sizes you need, plus several spares for the very fine ones that may occasionally snap.

Tapping a Hole

A tapping hole (read off a table) is drilled for a particular thread in the job. Using the tap-holder, held as vertically as possible, a 'first' tap is introduced. This is tapered towards the end to make it easy to introduce into the hole. Turn the tap clockwise. When you feel it start to bite, back off a quarter turn to clear the swarf from the flutes. Then turn a little further. Keep turning and backing off. If you are going right through a thin piece of metal, just keep feeding the tap through until its parallel sides are right through the metal and then unscrew it.

But if it is a blind hole, you will need to introduce a 'second' tap. This has a full thickness virtually to the end. Thus it will cut a full thread for nearly the entire depth of the hole. This takes care of most needs, but there could occasionally be a need to run the thread right to the bottom of a hole. For

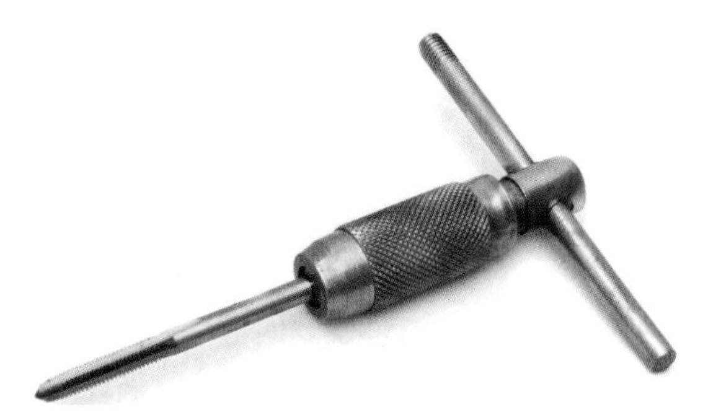

Smaller tap holders prevent too much force being applied.

this, a 'plug' tap is called for. However, you may go through your loco-building career without ever using one.

Taps have to be made of very tough material to be able to cut into steel. In small sizes they are particularly brittle and can snap easily, especially in very cold conditions. The secret is to keep backing the tap off, in the job, before it gets too tight. I like to hold small taps in a pin vice which I twirl between my fingers. This exerts less force than a big crossways tap holder.

USING DIES

To put a thread on the outside of a rod – and so make a screw – dies are used. These are round, with a small slot across half of one side. There are also two dimples on the outside. When a die is first introduced to the rod, the centre of three screws in the die stock is screwed into that slot. This holds the die open at its widest, while lightly cutting into the rod. You will see that the thread on one side of the die may be more tapered than the other. Use this tapered side face down.

When the thread has been lightly cut, withdraw the die and slacken off the centre screw. Tighten up the two outer screws. This has the effect of making the diameter of the thread slightly smaller. When the thread is cut a second time, it will be to its proper depth. If the thread needs to go right to the bottom of a shoulder on the rod, turn the die round so that the tapered side faces upwards, then make a third cut. For many small threads, with brass in particular, you can cut the thread in one pass. For

An adjustable die in a diestock – with a plain die for comparison.

a finer finish when cutting in steel, use a thread-cutting lubricant. There are proper lubricants to be had, but for small threads, a quick squirt of WD-40 will do the job perfectly well.

Where threads have to be really accurately cut at right angles to the job, there are various ways of achieving this. If a lathe is available, a tailstock die-holder grips the die. This is introduced to the job held in the chuck, then pulled around by hand. Likewise, a tap is held in the tailstock drill chuck. But there are some non-lathe methods available. The cheap bench drill will hold a tap true. A round rod could be held in the chuck whilst the die holder is kept flat on the table of the drill press. For larger threads, there is usually some dodge involving blocks of wood and G-cramps that will do the job: it just takes a bit of thinking out.

SELF-TAPPING SCREWS

Finally, mention should be made of self-tapping screws. They are anathema to traditional model engineers, but they work perfectly well in the low stress conditions of our smaller models. They are also perfectly adequate for joining thin metal to similar, or to brass or steel angle section. The only drawback is that they are not available with nice little hex heads. For a simple first locomotive, self-tappers have much to recommend them. The wrong screw heads will be invisible when the engine is steaming around the garden. Perfection can come later.

With these basic metalworking techniques at our disposal, we can already think about building steam locomotives, using some ready-made components, as the next two chapters will demonstrate.

A tender chassis being assembled with pop-rivets and self-tapping screws – where they won't be seen.

CHAPTER 4

Dempsey

This chapter describes a first project in detail and contains all the elements that go together to make a G scale gas-fired steam engine. A Ruby kit, by Accucraft Trains, USA, is a good hands-on way to get familiar with the workings of simple live steam. The mist clears rapidly. All of the engineering is done for you and it can be (nearly) bolted together straight out of the box. I say 'nearly' because there is a little bit of fettling and adjusting to be done as you go along. If you put together one of these kits, you will become familiar with many of the principles and components described elsewhere in this book. If you seek an easy life, you could do this conversion using a ready-to-run Ruby. The components are mostly just bolted together, and so can equally be unbolted. Putting a kit of separate parts together gives good insights into what is involved in scratch-building later on.

I have tried to describe the job in a way that will allow the beginner to get a successful outcome. So the more experienced modeller may find one or two of the techniques unusual. Although there is quite a change in the appearance of the loco, compared to an original Ruby, the actual new work is so simple that a drawing is not really needed.

Dempsey *Peter Jones 2007*

Ruby, as she was intended. Photo: courtesy of Accucraft Trains (USA)

A SIMPLE CONVERSION

As a first gentle exercise in hand metalwork, we are going to look at converting the US outline loco into an early British mineral engine. This type of machine is generically known as a box tank, for obvious reasons. The aim is not to model a specific loco and, it is true, we shall take a couple of liberties here and there in the cause of simple construction. But the end result will be an unusual live steam loco in its own right: one to which the builder will have contributed a considerable amount. A particular advantage of the Ruby design is that you have a working steam chassis and boiler – and hence a running engine. The original side tanks and cab are merely bolt-ons, thus helping to make life easy for us.

The frames are a simple-pattern American bar frame.

You will note that the engine has the spindly looking bar frames. This shows clearly the difference between common European/British practice and that of the USA. The plate frames of European practice make for strong construction, but they demand a well-laid track. To cope with the rougher nature of early US track-laying in particular, the bar frame could flex. Bar frames were used in limited circumstances in Europe, but the primitive box tank engine looks correct with them.

BUFFER BEAMS

The first step is a small one. The buffer beams need a nice little mod to make them appear right for an old British mineral railway look. Using a small nut driver, supplied with the kit, undo the two screws that are holding on a step assembly. Establish a nice little spares box; there will be a few choice items going in there. Start the collection by putting in a step. You will see that there are captive nuts recessed into the back of the buffer beam. Using the screws you have just removed, engage them into the nuts from behind and then pull free with a pair of pliers. The nuts and bolts

PROTOTYPE BOX TANKS

Although box tanks were made by several manufacturers, including Hawthorns and Andrew Barclay, the vast majority were built by Nielson. Indeed, they tend to be collectively known as Nielson box tanks. In the main, they were mostly simple four-wheel tank engines with wheelbases that were around 66in–80in. Wheel diameters seemed to vary between three and four feet. They first saw the light of day in the 1850s and were primitive by later standards. Early examples used an archaic Gab valve gear but this was soon replaced – mostly by Stephenson's gear.

Water feed was by a pump driven off a crosshead. This meant that water could only be pumped into the boiler when the wheels were revolving. In case of emergencies, the loco would be run along a siding until it reached buffer stops. The rails would be oiled and the wheels allowed to spin round whilst the engine went nowhere.

Very early examples had a saddle tank that left the smokebox exposed, but the classic box tank reached right to the front. The mudflap door, held up by a chain, was a very distinctive feature. Chimneys varied from austere to ugly, but at the back of the boiler you would often find a raised round casing with an elegant safety valve cover.

There would usually be a bucket of sand somewhere – for handraulic application. Coal was carried anywhere possible. It was common practice for the engine to have an improvised tender, consisting of a simple mineral wagon with one end knocked out. Brakes usually consisted of wood blocks applied to the back axle only by means of a lever.

The footplate side sheets had a distinctive curve and there was an unusual arrangement of a narrow back sheet. There were examples with a rather splendid toolbox on top of the tank – something we can make use of in the model. Although there were later refinements to box tanks, what has been described above is the classic outline fondly thought of by enthusiasts.

These engines appeared all over the country – partly because they were a comparatively cheap engine to buy new. There was a notable example at the Dalmellington Iron Company, but the most well-known examples seem to be on the Redruth and Chasewater Railway – with another somewhat refined variant on the West Somerset Minerals Railway.

They were supplied in various gauges, 3ft 6in and 4ft 0in in particular. Thus they lend themselves to a nice 'sit' in G scale/SM45.

are then screwed together; pop them into the spares box. Repeat with the other steps. In their place, we will put some dumb wood buffers. They can be flush with the buffer beam, as is the case with the model shown here, or they can reach higher, to be in line with the buffers of any rolling stock you have.

Although the prototype would probably have used oak for these, we will look for a nice piece of densely grained something or other. I have a few strips of wood that came from a demolished parquet floor. They are hard and well-seasoned, with a very pleasant colour to them. Cut your wood, with the grain going vertically, so that your blocks are accurately cut for width across the existing buffer beams. Remember that you have to span the pairs of holes that were used by the steps – and just a little bit more.

The buffer beams as supplied.

If you are making flush height buffers, as I did, don't worry about cutting the beams exactly to the height of the buffer beam. Cut them slightly oversize. Measure carefully to ensure that the buffers are all inset from the ends of the beams by the same amount (I used the tail of a vernier caliper). Superglue the buffers in place, making sure that the sides of each buffer are truly vertical but ignoring the fact that the top and bottom may be roughly cut and slightly oversize. Use clamps of some sort – even pegs – to hold the buffers tight to the beams and leave overnight. Superglue (cyanoacrylate) is notable for its instant setting, but it doesn't achieve full strength immediately.

The steps removed and the oversize buffer blocks glued on.

When tomorrow comes, run a small drill through the recesses in the backs of the buffer beams and into the buffers. But take great care not to drill right the way through. This pilot hole will let you screw some tiny roundheaded woodscrews in from the back, to reinforce the superglue. Those little domed heads should neatly sink into the recesses that were drilled to take the buffer beams.

Now file and sand the excess height of those buffer heads down, until they are flush with the beams. It is easier to produce an accurate overall finish by doing it this way. And if you have a belt sander, then a minute or so of the top and bottom of a beam assembly being touched against that will produce instant perfection.

Finally, I rather like the idea of giving everything a coat of matt varnish. There is one extra refinement if you wish – cut four rectangles of thin mild steel the same size as the buffer faces and glue them on. Iron-faced dumb buffers were harder wearing, although not universally applied. There is something rather nice in a model to see that natural wood and steel together. An expert would also use countersunk pins to hold the faces in place, but maybe we don't want to get too ambitious just yet.

STARTING ON THE FRAMES

Whilst that superglue is hardening on the buffer beams, we can turn our attention to making a start on the chassis. Ignore the instruction to bolt angle brackets to the frames. We will not need these because the side tanks will not be used in this conversion. The first move is to fit the bushes on the axles of the wheel sets into the two frames. Double-check that the frames are the right way round from the instructions and that we are putting the axle with the eccentrics into the rear locating holes. Those instructions will also tell you to note little flats on the bushes and that these should locate against flats in the bottom of the holes in the frames. These bushes should push smoothly sideways into the frame openings. Force should not be used for fear of bending the frames. Mine wouldn't go in at all. The openings in the frames had to be slowly enlarged with a round file and a flat needle file until they did. If this happens to you, just patiently file away at each opening in turn until it is a gentle friction fit.

The front and rear frame stays can be screwed in. When the wheels are in place, you should now have a freely rolling basic chassis. It is possible that the wheels might be slightly stiff to turn; if only marginally so, not to worry. But if the fit of those axles in the bushes is a bit too tight, then we need to ease it and bed things in. My favourite method is to find a belt drive somewhere. The bench drill with the cover hinged up is fine. Set the bench drill running and then gently push one of the wheels against the belt. This will cause the axle to revolve at high speed. Dribble a drop of oil onto the axles and spin once again. In the unlikely event that wheels may seem *very* stiff to turn,

REGAUGING A RUBY

This is not for beginners, but, if I wanted to regauge a Ruby to 32mm gauge, I would change the long chassis into three open rectangles. The area around the driving wheels would be narrowed and the length of the rocking arms reduced accordingly. The frames fore and aft of this middle 'box' would be left at normal width to allow most other components to bolt on as per the instructions. The assembly of valve chests (three blocks connected by pipework) would need to have those pipes shortened so as to align the cylinders with the connecting rods.

I make no claims that the above is the only way of doing the job. but it shows the sort of thinking that will be required when you are further on down the road.

The front wheel set, including the bushes for mounting in the frames.

The rear wheel set with eccentrics and inside rods already machined.

The coupling rods are simple blanks, but have a reasonable profile.

undo the parts you have bolted together and screw them in again. Ruby wheels have a good name for free running.

The frame should roll smoothly on its wheels and also all four wheels should sit smoothly on a true flat surface, such as a piece of plate glass, without any trace of rock. Ruby kits are generally quite good in this regard. But if you can detect a very faint rock, partly slacken off the screws to the frame stays. Take courage and frames in your hands and give a twist in the opposite direction. Tighten the frame stay screws and see if it is now flat. Sometimes you can be lucky and get a small twist in a frame trued up first time. If not, repeat the process until all is well.

I like to keep a rolling chassis pressed onto a piece of foam, in order to stop it rolling off the workbench.

The two valve chests and a centre valve block come ready assembled.

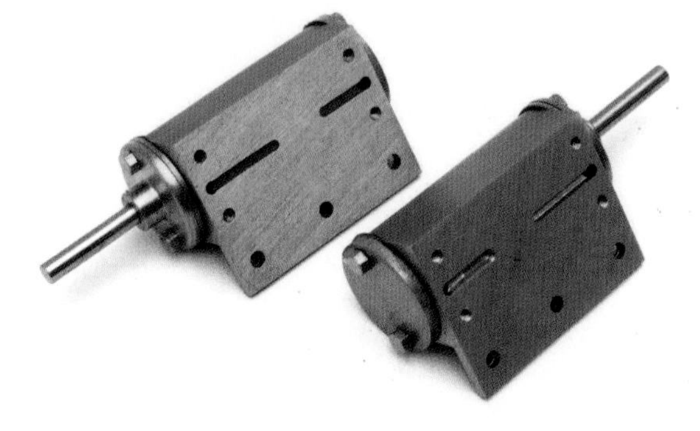

Steamways are machined into the cylinder faces.

The rocker arms are simple – worth bearing in mind for future projects.

The two cylinders will bolt onto the valve chests.

THE REAR FOOTPLATE

By now, those buffer beams will have set hard and you will have added those tiny woodscrews. A coat of matt or satin varnish is applied to one side of each assembly and then left to dry fully. Then the other sides of the two beam assemblies can be varnished. This means we have time on our hands and can get on with something else.

Filing the footplate around a washer. The modified steps can also be seen.

The rear footplate is a large flat item with turned-up flanges at either side. We want to make a couple of mods here. You remember those steps we took off the buffer beams? We are going to reuse two of them to make diminutive stirrups steps under the foot-plate. With each one, bend the arms inwards to a right angle, as shown in the photograph. The U-shape should have parallel sides so that the steps will hang down about 12mm. Don't try to make the bend a neat hard right angle, as there is a risk that the brass could snap. The slightly curved bend will be hidden under the footplate – so only you will know.

Offer up the arms of the steps so that the steps will line up with the back corners of the plate. With a piece of sharpened rod that has a small right angle at the end, scratch the location of the fixing holes. Drill these through 2mm, then put the bent steps safely in store for now. Make a note: the front screw on the left side would stick up and stop the gas tank sitting properly. You will therefore need to find a 2mm countersunk screw to replace the existing hex head and the hole will need to be countersunk. I gently kiss the hole with a slowly revolving larger drill to form this.

ROUNDING OFF THE FOOTPLATE

The front two corners of the footplate need to have an 18mm radius put on them. Fortunately, there is an easy way to achieve this. It merely needs a 36mm diameter steel washer. Clamp this to the corner of the plate so that it sits level with the front and the side. You can saw away the bulk of the metal in straight lines (I used a grinding wheel) and then finish to the radius, with the job held in the vice, with a couple of files. Finish off with a bit of emery paper wrapped around a flat stick.

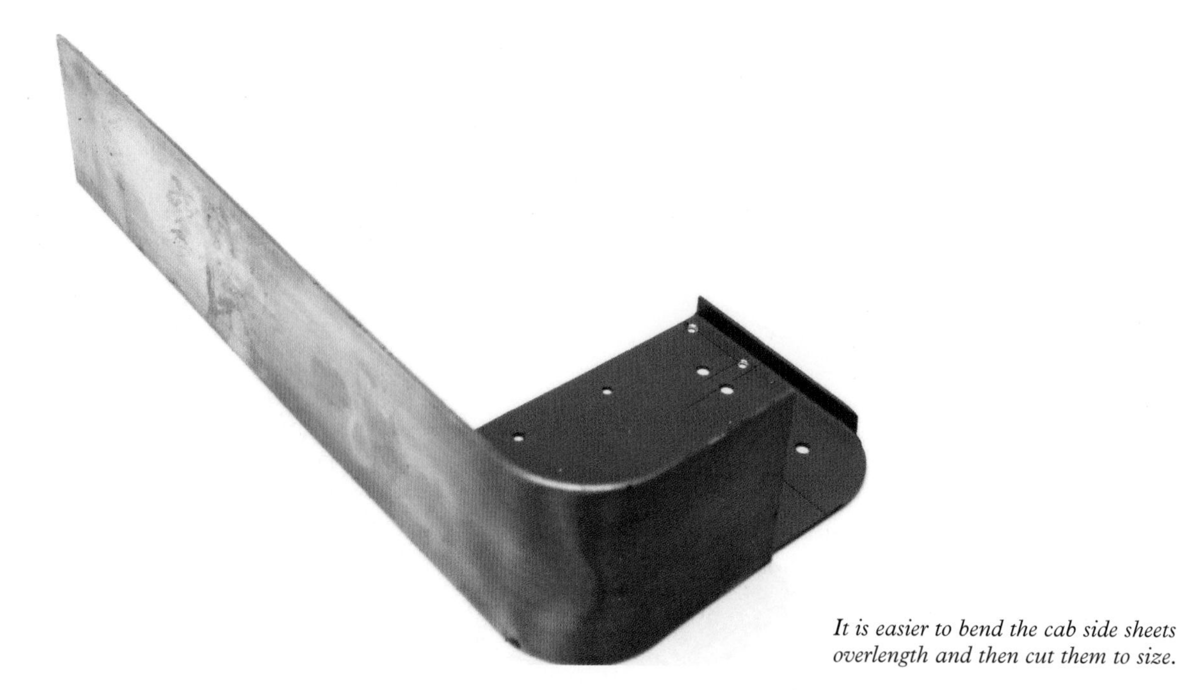

It is easier to bend the cab side sheets overlength and then cut them to size.

As all this metal is falling away, you will be waving goodbye to the twelve-month warranty issued by Accucraft Trains. But be of good cheer, as your eventual model will be a machine to be proud of.

THE CAB SIDE SHEETS

These are made from fairly thick sheet brass – about 40thou. You are looking for a single piece of brass about 30cm long. If you can get it 50mm (2in) wide, that will save having to make a long, accurate cut in a larger sheet. (If you do need to make that cut, and you are new to this, the usual rules apply – cut slightly oversize and file the last little bit down to a straight piece of steel clamped to it.)

The task now at hand is to put an 18mm bend in the brass so that the side sheet will match the curve in the front of the footplate. Don't try to cut exactly to length and then bend. Do it the easy way. Put a bend in the full length of the strip, inboard from one end. However, before we do that, the brass needs to be softened. Lay it flat on the hearth and get a good roaring blowlamp to work, playing it on the area to be bent, which needs to glow a dull red. Leave it to cool naturally.

To put the bend in, we need a former of some sort. A piece of steel bar or heavy steel tube is what we want. Nominally this should be 36mm diameter, to match the washer we used to file out the curves in the footplate. But there is usually a bit of 'spring-back' after bending. So look for a bar that is slightly smaller – around 32–33mm. Put some cardboard liners in the vice jaws and then you have the tricky task of gripping the steel bar and the brass strip in the vice, at the same time. It is better done with two pairs of hands, although with a bit of juggling you can do it on your own. The bar goes horizontally crossways and the strip must stick up exactly vertically. Check it with a square. You can see now why it is easier to put a bend in an approximate position rather than an exact one. With the pressure from the vice very slightly loosened, you can make final adjustments for squareness with a couple of light taps, before retightening.

Take an offcut of timber – say, 2in × 1in – and use it to bend the brass strip down towards you. It will go easily. Peer at the job end on to check that you have bent a decent 90 degrees. Take the brass out of the vice and offer it up to the side of the footplate. With luck, it will fit round the radius exactly. If not, the brass will still be soft enough to bend around a former by hand, or even just in the fingers.

The smokebox saddle is machined from a block of brass.

Cut the side sheet to length at both ends. You will have a short offcut of brass, which will go into the 'useful brass bits' box. The remaining length will now be used for the other side sheet; repeat the procedure. After this, there will be a longer offcut as well, which we will use for the cab back sheet later. Many box tanks had extra pieces of side sheeting bent inwards towards the centre of the cab, but our model has to host a gas tank, a reversing lever and a lubricator. So we will leave these off.

FITTING THE SIDE SHEETS

These will be soft soldered to the flange and the front edge of the footplate. But Ruby is a compact engine and can get very hot, so, just to be on the safe side, we will add a couple of hex-headed screws and nuts on each side. Sizes are not desperately important. The loco uses a lot of 2mm screws and the kit supplies a nut runner for these, so that seems as good a size as any.

Life may be easier if you soft solder a side sheet just to the flange, leaving the front curve untouched, as a starting point. This means that the two parts are held accurately together whilst you drill a couple of 2mm holes. Put the screws and nuts in tightly. Paint a little flux onto the curved edge of the footplate and inside the curve of the side sheet. Clamp these tightly together. I have a large vice that will accommodate such things, but if you haven't there is always some way of clamping – even if it means knocking up a jig out of scrapwood.

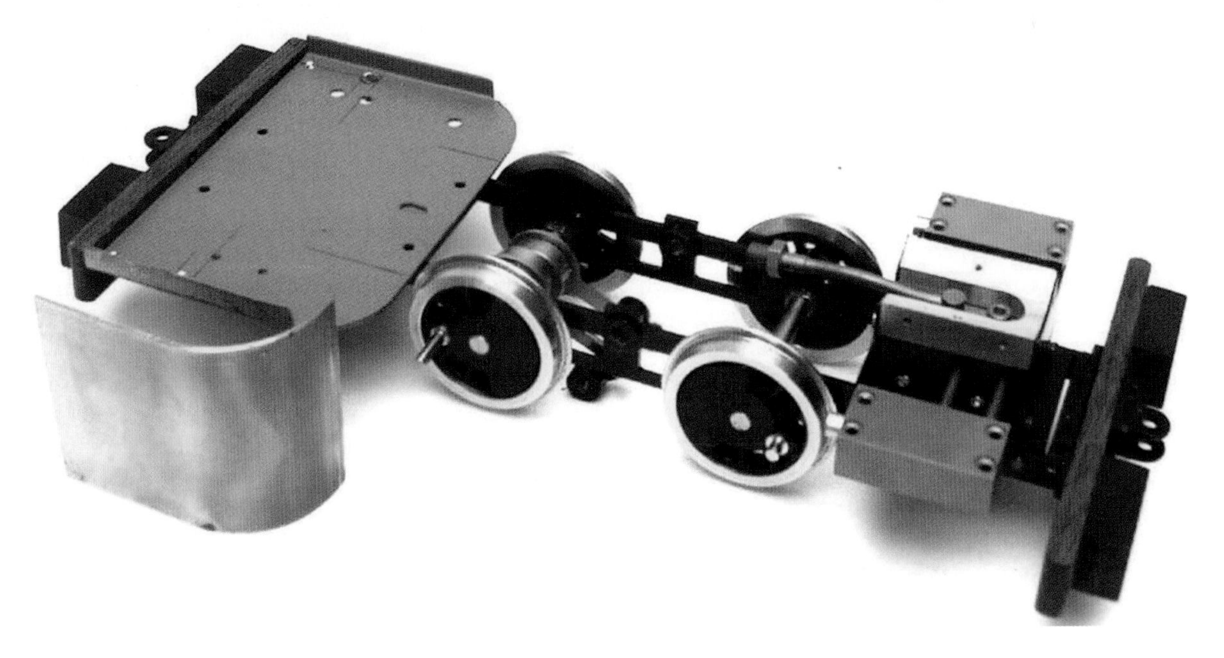

A cab side sheet has been cut to length.

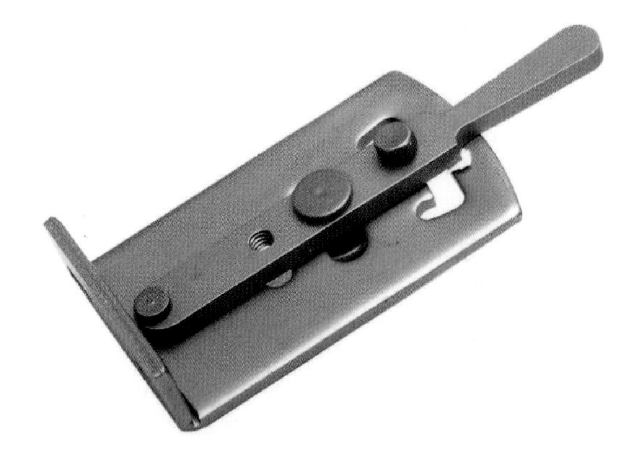

Accucraft offers a simple reversing lever – again, worth remembering.

Both side sheets are in place; motion and cylinders fitted.

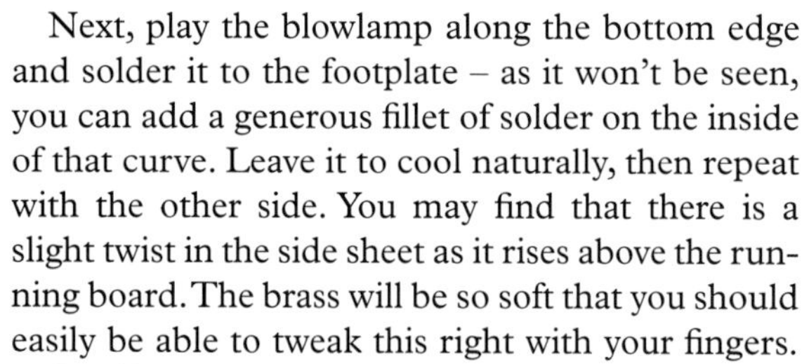

Next, play the blowlamp along the bottom edge and solder it to the footplate – as it won't be seen, you can add a generous fillet of solder on the inside of that curve. Leave it to cool naturally, then repeat with the other side. You may find that there is a slight twist in the side sheet as it rises above the running board. The brass will be so soft that you should easily be able to tweak this right with your fingers.

At this stage, I also soldered a couple of brass number plates onto the cab side sheets. Given a choice, it is always a better bet than gluing with Araldite. Finally, we can now bolt those steps in place permanently.

Time to treat ourselves to the sight of the project standing out on the track.

FIRST PAINTING

Run a green scouring pad lightly over any shiny brass areas, just to provide a key for the paint, and then give the assembly a wash in warm soapy water, scrubbing with a toothbrush to remove any trace of flux or greasy fingermarks. Leave it to dry under a cover to keep off as much dust as possible. You may wish to spray on a coat of brass etch primer, although I usually find that the light scouring is enough.

If you are going to paint your loco gloss black, give the assembly four or five thin coats of gloss black, using a cellulose car-paint spray. Leave several hours between coats, storing under a cover to prevent dust. When you have built up a good gloss finish to match the existing Ruby paint, put the assembly aside for several days. It still wants to be in a dust-free environment and certainly nowhere damp. The steady heat emanating from the top of a radiator is particularly kind to paint finishes.

My Ruby was to be painted in a very dark, workmanlike green, although at this stage I merely gave the assembly a few coats of satin black – including the number plates. When this was completely dry, I cleaned off the raised parts of the number plates with an emery stick (an old track rubber can also be used). The dust was blown away and the job was done. At this stage, we can treat ourselves to the spectacle of our model, part-built, sitting out on the track. We are making progress.

VALVE GEAR

Normally, this would be a complicated part of constructing a small steam engine. However, one of the reasons that I chose this kit, as a learning aid, is that the valve gear is simply described. By following the words and drawings, you will have assembled everything and set the timing in about an hour – and clearly understand how it works.

Along with several friends, I had difficulty fitting the tiny e-rings to the valve rod pins. In the end, I threaded them 7BA (2mm would be just as good) and ran brass nuts onto them. When the engine had proved itself to be a good runner, I put a whiff of soft solder, using a soldering iron, to fix the outer nut to the thread.

At this stage, everything turns and the valves go in and out, but it will all seem horribly stiff. This is natural. There needs to be a running-in phase. If you have access to a garage air line or a compressor, fake up an adaptor that will screw into the steam pipe coming from the cylinders. Put oil everywhere! Set the air line going and the moving parts should revolve backwards and forwards on command. There is a good chance that the timing will need adjusting slightly, but, in conjunction with the original Ruby instructions, you will be familiar with the process by now.

If you can press a driving wheel against some sort of rubber tyre that is revolving steadily, this will turn the engine over mechanically. Keep applying plenty of oil. Many years ago I used to use a 100mm toy pram wheel that was turned by hand by someone else. It was a lash-up but it worked.

THE BOILER

Our intended conversion calls for a couple of preliminary mods to be made. Don't worry; they are very simple. The smokebox door is unscrewed and goes into the spares box; file or grind away the little hinge and locking points. The front of the tank assembly will go over the opening, but will be subject to considerable heat from the gas flame. To overcome this, we will cut a disc of a stiff, heat-resistant material. In the past I could advise using asbestos, but this is, rightly, taboo now. Fortunately, there are substitutes. I had a small pad of a modern substitute, about 4mm thick, which had been sold with a soldering iron as a heat-resistant material. I cut my disc from this, marking it out by pressing the front of the boiler hard into it, which left a perfect indentation. I was able to cut it out using repeated gentle cuts with a scalpel. It was cut very slightly oversize and then gently trimmed until it was a hard push-fit.

The boiler as supplied. You can see the opening for the fire tube on the right. The method of fixing brass boiler bands is also made plain.

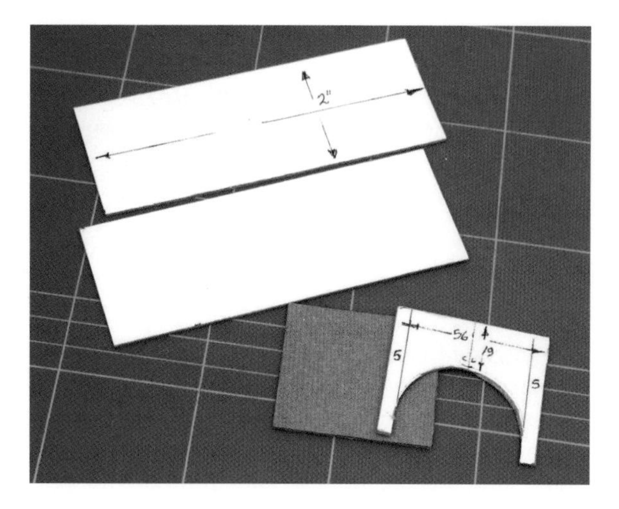

Making a mock-up of the box tank in card.

The boiler from the smokebox end. The door will be discarded.

The dummy tank is propped in place.

The regulator has been screwed to the boiler and the dummy dome is offered up.

REPLACEMENT CHIMNEY

For a simple life we could use the existing chimney, in an arrangement to be described when we come to fabricating the tank. However, to my eyes, the utilitarian straight chimney was a feature of many of this type of engine and I would like to capture that. For the purposes of this chapter we will do it without recourse to a lathe.

You will see that the arrangement features a hollow screw sticking upwards from inside the smokebox, using a shaped washer. The chimney screws down onto it. If that shaped washer is omitted, the hollow screw projects further upwards. To turn this into a shaped nut, 6mm is cut off the bottom of the chimney. This then screws onto the hollow tube, as shown. The remains of the original chimney will go into the spares box, of course. It is through this short locating stub that we will light the engine.

Our plain chimney is a piece of 15mm copper tube. It needs to be able to slip loosely over the projecting part of that screw. Cut it off somewhat

The safety valve is small and neat. It is almost a pity to hide it.

When we say that the hole in a gas jet is small, we mean it!

The gas tank assembly has a huge 'safe' control knob, as supplied.

Ruby has a typical displacement lubricator in the cab area.

oversize for now. When the tank is finished, the tube will be cut to a set distance above the tank top. Incidentally, copper tube of this size is most easily cut using a cheap pipe cutter. When the time comes, we could leave the top of the chimney completely plain, but many of these engines featured a chimney with a small lip. Between now and when we are near to completion of this engine, keep your eyes open for a small ring of some sort. One possible source is to use a suitably sized spectacle ring, sold for fitting to locos that have round openings (spectacles) in the cab front and back. Over the years, I have gradually filled an old tobacco tin with rings that have come my way. A further possible source is porthole rings from a model boat supplier.

A filler bush can be drilled through and tapped to take an air line for testing purposes. The threaded hole would be blanked off with a screw and O-ring.

CONTROL KNOBS

The control handles of the throttle and the gas valve may be fitted with large brass levers or big moulded-on plastic knobs. Both look a bit crude, but they are there to help protect your fingertips from getting burnt. With years of running steam engines, one develops hardened fingertips, which seem impervious. You could unsweat the levers or carefully saw away the plastic part of the knobs. This leaves a neat brass bush, which you can fit with an extended grub screw, as an indicator. This looks far less obtrusive. *But do not do this if children might run the loco!*

The chimney bush will be modified to take a plain replacement chimney.

It is worth examining the regulator in detail to see how it works.

THE ORNATE DOME

One characteristic of box tanks was an interesting dome, immediately aft of the tank. The designs seem to vary considerably, but they had the common feature of a plain-looking casing, upon which sat an ornate brass safety valve cover. We will attempt to model this. There is just room between the tank and the Accucraft regulator to fit one.

In keeping with the spirit of this chapter, we will make one without recourse to a lathe. Start by cutting a 20mm length of 22mm domestic copper pipe. One end needs to be shaped to follow the curve of the boiler snugly. The secret of doing this neatly by hand is to file rough depressions to match that curve, using a half-round file. Next, find a round bar of something, such as plastic tube or turned wood, that is a slightly smaller diameter than the boiler. Wrap a piece of emery paper round it, rough side outwards, and grip in the vice. Then push the copper tube to and fro along it. Copper being so soft, it will adopt the correct profile in a few minutes.

There needs to be a top to this casing. If you do not happen to have a brass disc or copper washer that is the right diameter, cut a disc of copper roughly – slightly oversize – and soft solder it to the top of the copper pipe. File it round by hand until it sits flush on top. For the ornate safety valve cover, you should rummage through your workshop until you find something that looks about right. Perhaps it will be a section cut off an old brass toasting fork, or the neck of a small copper ornament; there were different shapes of cover on the engine prototypes, so you are bound to find something that looks the part. Solder this on the assembly. Scrub any flux off and put it to one side for later. For fixing it down, we are going to be sneaky. It will be bolted with a spacer to the back of the removable tank on top of the boiler. It is as simple as that.

Parts for the dummy dome.

The tank being assembled in a soldering jig.

The tank back sheet.

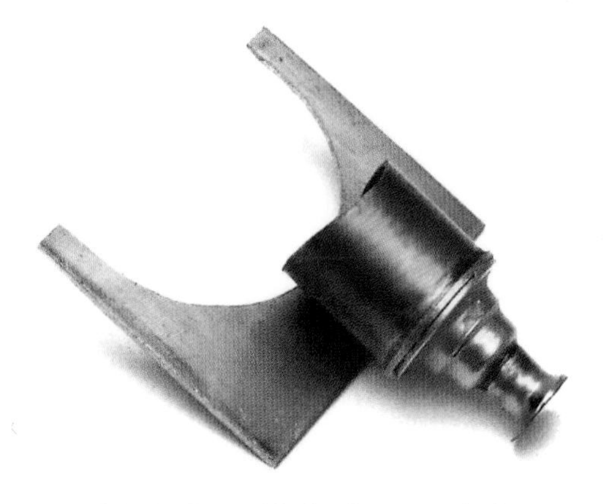

The tank back sheet with the dome attached.

BUILDING THE TANK

The tank is made from five major pieces of 1mm brass. The sides are plain rectangles 132mm long. For an easy life, I have suggested that they are made from 2in wide strip, to save cutting accurately to width. But the tank will have a nicer 'sit' to it if the sides are 46mm high. If you plan to fit name plates and prefer to bolt them in position, line them up centrally and superglue them in place. Then drill through at either end for bolt holes. Turn the side upside down, then play the flame from a blowlamp on the back for a minute; the plate will fall off. Just

CARD MOCK-UPS

A useful tool for the beginner to use, when making new modifications or conversions, is the card mock-up. For example, to avoid making mistakes in brass, it is a good idea to fabricate a dummy version in card and stripwood. This principle is especially useful when you are working on a design of your own, such as a saddle tank or new cab.

A centre line was drawn along the top of the card tank. I could then mark where the chimney needed to be and where the safety valve should project. The tank was built as an open rectangle at first, with the back end cut out to suit the diameter of the boiler. The front end was merely a plain rectangle lightly glued in place. The tank could now be dropped in place and the front end propped up so that the top of the tank was level. I could now measure up for the full depth of the front. The temporary front was removed and the full deep front (still in card) was glued in place. So now I had a tank that was supported at the front and back ends. I noted that I had allowed the sides to fit outside the ends and had made allowances for the thickness of the sides. Finally, when I was satisfied that all was as it should be, I could start to make the tank in brass.

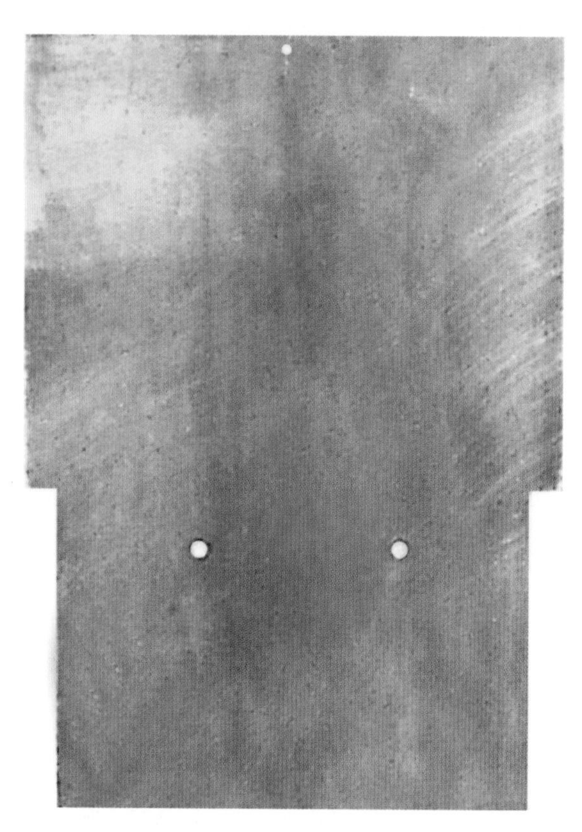

The tank front.

The tank front with detail overlay.

to be on the safe side, make a single punch mark on the back of the plate and on the back of the tank side. When you repeat the procedure on the other side sheet, put two little punch marks on the back of the sheet and the name plate. This is just in case your holes aren't drilled in exactly the same place on both sides. By putting the right plate with the right side, you know it will go back in exactly the right place.

The front plate is a rectangle of brass 80 × 57mm. It has two notches cut out 33mm high × 3mm wide. A small hole is drilled near the top, in the centre, to accommodate some fine chain (start scouring your scrap boxes for this, in anticipation).

There is some simple detail to be built up on this. To make life easy, I suggest making this a separate layer. It consists of another piece of brass that is 57mm high × 44mm wide. The top is rounded off to a semicircle of 22mm radius. I suggest marking this out and then sawing away most of the surplus metal with a hacksaw. File it bit by bit down to the final curve. If you come across a washer 44mm diameter, then file it down to that. On this additional layer I have shown a bit of rudimentary detail. There is a rectangular plate 34 × 20mm soft-soldered in place. A 50mm length of 2mm brass rod is soldered above it. You can put more accurate detail on this front if you wish, but the engine will look quite happy in the garden as described. A couple of holes are drilled through the door, plate and backplate. Later on, the detailing plate will be bolted to the main front plate.

ASSEMBLING THE TANK

You would think that assembling a simple rectangular box out of four pieces of brass is about as easy as it gets. But it is that very simplicity that shows up the slightest error, so I am going to describe a method that produces a reliable result. We will make up a jig from scraps of wood. For the base, an offcut of white-faced chipboard is as good as any. The sides of an open box will be made out of 2 × 1in planed timer (check that you have a proper right angle between two faces). Draw a pencil mark across the base and then offer up a length of wall against it, gluing with a few drops of superglue. Push two thicknesses of brass against this and then place the rear end at right angles to it. This gives you the total width of the inside face of the jig. Make a pencil mark and then slide the assembly sideways and make further marks. Joint them with another pencil line and glue a second side wall in place. You need to cut two end pieces for the wall very accurately, with truly cut ends. They will be a tight push-fit between the long sides, at exactly 90 degrees. In effect, what we have done is to create an enclosed space into which the two sides will press against the insides of the long walls – held fairly tightly in place by the two ends fitted inside the two walls.

When all four walls have been assembled and you are happy, turn the board upside down and pre-drill a few holes for some woodscrews to go up into the walls. The jig is nearly finished. Turn it back the right way up and put the four parts of the tank back in place. Cut a piece of 3mm ply to fit tightly inside the four brass pieces and press down to the floor of the jig. Because the tank is being assembled upside down this means that the top of the tank will be recessed inside the sides as per the prototype. Hopefully everything will be a stiff press fit. Remove the four brass parts of the tank and lay them out logically.

Cut four pieces of small brass angle the right length to reach from the top of that extra thickening piece of wood to the height of the tank – in other words, to fill the full height of the inside of the finished tank with its top in place. Also cut two further pieces of angle that will run on the inside of the side tanks from one piece of corner angle to the other end. Now thoroughly 'tin' all the faces of the inside of the tank and the outside faces of all angle pieces. Put the four sides back into the jig. The thickness of the solder may make them especially tight for now.

Now for the sneaky bit. You will need two 'braces' that will go from the inside corner to the diagonally opposite inside corner. They are best made from stiff wire or similar – they need to be rigid but with a bit of flexibility to them. They will hold the corner angles in place, pressing into them to sink finally into their respective corners when the blowlamp is played on them. Incidentally, when you are using a blowlamp in this enclosed space, it is easy to run out of oxygen and the flame will go out. But you will soon get the feel of just the right place to hold the lamp.

You are, of course, aiming to get all four corner angles securely soldered in place. With the diagonal braces holding them tight, you could add a bit more solder to the edges of the angle pieces. It can't be seen from outside the tank, so only you will know. Drop the lengthways angles in place and soft solder them. By now, the jig will be scorched and battered, but it has done its job. Go and have a cup of tea and let everything cool down naturally. When you come back, it might be quite hard to get the tank assembly out. If it is really tough, drill a ½in hole through

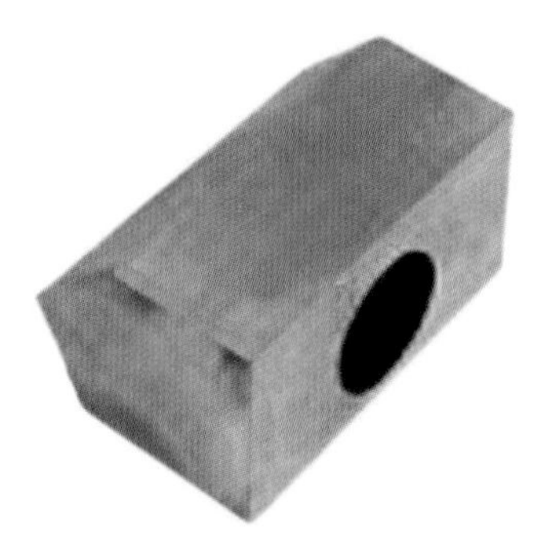

The toolbox cut from solid. Note the recess to accommodate the safety valve.

from the underside of the baseboard and, using a stub of wood dowel, tap it out from underneath. You should now have the sides of the tank rigidly and accurately soldered together. With all that heat and flux, the brass will look horribly scruffy, but ignore its appearance for now.

Go round the tank, drilling and tapping M2 ('metric size – 2mm') so that you can reinforce the soft solder with some little hex-head screws. You will have some left over from unwanted parts of the original kit, although you may have to buy a few more. The soft solder on its own will probably hold everything in place, but if the boiler were to run dry, it is just possible that the loco could overheat enough to melt a joint or two. Having a few screws in place will hold everything together whilst it cools down.

A small toolbox is sawn and filed from an offcut of brass (or aluminium) bar. Cut out a rectangle of brass to drop into the tank top, landing on the angle. Carefully mark out the opening for the chimney to pass through, as well as the opening for the safety valve to poke up. Stick the toolbox in place with superglue, as well as whatever you are using for a tank filler cap. Drill through and tap M2 from the underside to hold both these items in place; two screws for the toolbox and one for the cap. Heat from underneath to free these two items.

TANK TOP FITTINGS

The chimney tube should be pushed down through the tank top until it locates on the stubby remains of the original chimney. Mark where the tank top comes on the tube and slide some small brass ring or fitting – whatever you have in the scrapbox – to the mark. When satisfied that all is well, withdraw the tube and tack-solder the ring in place for now. Whatever you have found to make a simple chimney cap can now be slid onto the tube. I am going to suggest that the top of the chimney should be 45–50mm above the top of the tank. Cut the tube to length. You could actually leave the top of the chimney completely plain, in the manner of some of the prototype box tanks, but this really does look a bit spartan to my eyes.

Clean the dregs of glue off the underside of the toolbox and the tank below it. What we need to do is screw the box down, but with a gasket underneath it. That hole in the underside of the box acts as a receiving chamber to let any escaping steam be directed down to the underside of the tank (which has no bottom anyway). A few wisps of steam between the frames looks quite acceptable – a jet of it shooting out from under the toolbox doesn't! The gasket is a plain, thin sheet, with a hole cut in the centre to accept the safety valve loosely.

A gasket can be made from thin steam gasket sheet if you can get some easily, but an alternative is to make one from a piece of stout brown paper soaked in oil. In my case, I used a liquid gasket, sold in tubes from a motor accessory supplier. Solder the chimney and any rings in place on the tank top. Screw and solder the tank filler cap in place. Finally, screw the toolbox in place, using whatever gasket material you have chosen.

This is now a good time to check that the tank sits level on the boiler. If you peer underneath at the bottom of the tank sides and the bottom of the boiler, they should look parallel to the naked eye. You may need to do a little filing at the front or back end if not.

Two little extra plates will be needed to suggest the sides of the smokebox projecting down from under the tank. These fit in over the top of the cylinders and down into the gap between them and the front buffer beam. The exact dimensions may vary, depending on the size and location of the box tank itself. It is therefore best to make a cardboard template first, adjusting the size and shape until you are satisfied. Finally, cut them out from brass angle – fitting outside the front plate – and soft solder.

This now concludes the making of the tank. Using gentle abrasion, such as a scouring pad or a brass (not steel) wire disc in the bench drill, clean away all the disfiguring heat marks and any surplus solder. Pay attention to the corners of the tank being neat. With a scouring pad, or very fine wet & dry paper, remove the shininess from all the metal. You don't want deep scratches – just an even satin finish. Then put the parts into a solution of two dental cleaning tablets and leave to soak overnight. This removes all the grease and grime. In the morning, the water will look slightly blue.

FINAL ASSEMBLY

That is all the hard work over and done with; time to put things together. Make sure that the heat-resistant front to the smokebox is pushed in place (or that a scrap of heat-resistant cloth is glued to the inside of the tank front). Screw the copper exhaust pipe down into the valve chest, as per the instructions. You will notice that this pipe is crimped at the top, with a little slot cut in the side, just underneath. This came about through experience over the years. It stops jets of boiling water shooting out of the top of the chimney when starting the engine from cold. This makes a mess of the engine and surrounding track. On the other hand, it means that the steam exhaust is less clean later and that water always seems to be dripping down onto the valve chest and spitting/coughing. So I am in a minority that prefers that the exhaust is just a straight jet. When starting an engine, I drop a plumbing copper elbow temporarily over the chimney to direct any hot water sideways. If I am really fussy, I replace the steam pipe with a longer one that has a loop bent in it; this briefly visits the fire tube in the boiler, making the exhaust steam much drier. But it doesn't make any difference to how well it all runs, so by all means stick to the existing arrangement if you prefer.

Using a spacer, bolt the rear dome to the back of the tank. Screw the extra layer of the smokebox door onto the tank front. Drop this into place on the boiler, to see that all is okay. Then drop the tank top in place. At last . . . we have an idea of how things are looking. Most encouraging. Give the engine a couple of trial runs. Despite your earlier running in, it may well still be quite stiff, but that will pass.

After finishing off and painting – suddenly we have an engine. I like to see two crew on a footplate.

The cab area now has a wooden plank floor; the controls have also been cut down to size.

The front end is neat and simple.

When a loco has an optional tender, that cab back sheet should be removed.

PAINTING

If all is well, partly dismantle the engine again and make it really clean and grease-free. I would be inclined to put the box tank assembly back into the denture tabs solution for another soak. Spraying raw brass with a self-etch primer (try Phoenix Paints) does make for a better bond. Then spray the tank, the tank top, the bottom of that rear dome and the cab side sheets in body colour. The side sheets can have their surroundings masked off with pieces of plastic shopping bag and masking tape. Spray-paint by putting on lots of very thin coats, leaving time for them to harden in a warm, dry atmosphere, while protecting them from dust. Do remember that, just because a sprayed paint is

OPTIONAL TENDER

It is a simple job to put together an appropriate tender. I folded up a couple of examples in card first, just to be satisfied with proportions and to try to capture that primitive wagon look. The floor used sawn stripwood, with four axle box castings from the spares box, pinned and glued in place. The rest was just assembled from ice-lolly sticks and odd scraps, roughened slightly to give an impression of coarse grain. I sprayed the finished body in the same paint as the loco and enjoyed seeing the roughness of the wood texture. The load of coal was a roughly shaped triangular wedge of wood with the top hacked at with a small chisel. It was sprayed satin black and then bits of coal and coal dust were sprinkled onto some wet glue on top. A similar tender could also conceal a radio receiver and servos.

. . . as does the tender back.

The tender side features deliberately rough wood . . .

From the front, you can't see that the coal is just a thin layer glued to a roughly shaped block of wood.

touch dry, it may not be hard for some days. Be patient. You have worked hard enough to get to this stage. Don't spoil it by impatience now.

The extra layer on the tank front can be sprayed satin black, as can the 'smokebox sides' and the inside of the cab back sheet. The chimney could be sprayed with a heat-resistant black paint, leaving any shiny cap polished. But, in truth, I usually get by with brushed-on satin black enamel.

With cellulose car sprays in particular, the received wisdom is that the paint is left for a week to harden and is then 'cut back' with T-cut or a similar product to produce a high gloss. Speaking personally, I don't like this as I feel the result looks somewhat toy-like. I prefer the hint of a satin finish, so I leave it slightly dulled.

The painting style of these engines was often very plain and functional; no lining, often not even red buffer beams. In my opinion, a fairly sombre colour suits these engines. I used a Ford Pine Green and satin black.

FINAL DETAILS

We can add that little bit of chain that hangs down from the smokebox front (to hold the flap open when required). Coupling pins are simply cut-down pop-rivets. I added a wood planking floor to the cab area, but it is not essential. Screw any name plates or number plates in place. The job is finished.

A piece of Veroboard, marked out for pilot-drilling rivet holes accurately.

MORE ADVANCED REFINEMENTS

This project has been designed to make the principles and the hardware of a small steam engine familiar. Along the way, we have built an original-looking engine without recourse to advanced model engineering practice. However, there are more advanced refinements that could be added. For example, the prototype dripped with rivets. In those primitive days, riveting small plates was the only option. If these are to be reproduced, then they *must* be accurately aligned and spaced; nothing looks worse than a wobbly line of rivets. Fortunately, using electronic Veroboard can help us to achieve these straight lines (*see* Chapter 3, 'Riveting' for further details). Use conventional riveting techniques to fit the angles and tank components together in the proper manner.

With access to a lathe, you would have more control over the shape of things like the chimney cap and the dome casing. Chemically blackening the pipework and the regulator/lubricator would tone things down. Cylinder covers could be made and painted (hot brass cylinders are notoriously poor at

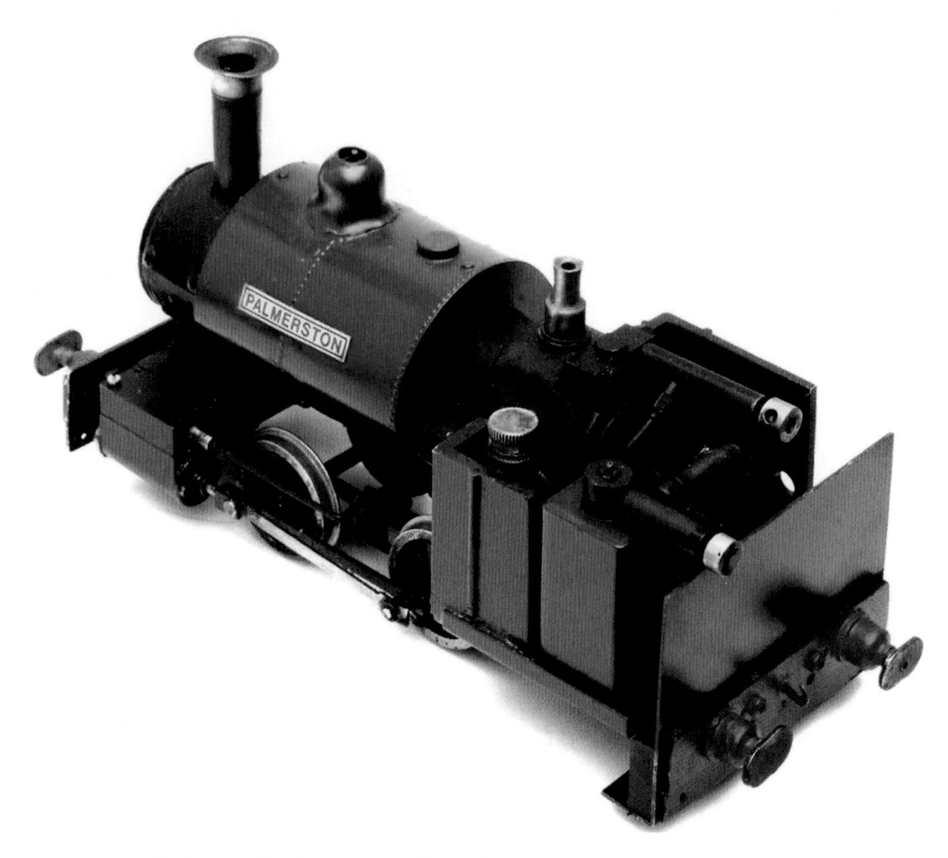

An earlier Ruby conversion of mine . . . this time to a saddle tank.

retaining ordinary paint). There were also some distinctive curved pipes that went from the cylinders into the smokebox. These could be soldered to the cylinder covers. Indeed, a lot of the assembly could be done using silver solder instead of soft solder and reinforcing screws.

Although I have described a Nielson box tank, there are some other alternatives that would make splendid conversions. The early Ffestiniog box tanks/saddle tanks would use various similar techniques and design ideas to those described above. Ideally, they should be re-gauged in to SM32 to get the right appearance. Another alternative is that the basic Ruby working parts can just about be contained in a standard gauge G1 conversion – at 10mm to the foot. I would be inclined to lower the boiler and shorten the smokebox. The footplate should be narrowed until it is in line with the width of the cylinders. This will call for the gas tank to be relocated. Other than that, you have all the working parts of a steam engine, fitted inside the envelope of a Gauge 1 loading gauge. Putting a tram engine body over the basic Ruby, instead of its original cab and side tanks, is a simple proposition, mostly using rectangles of brass and brass angle.

So many options . . . and they are all there waiting to be explored.

CHAPTER 5

The DACRE Principle

Long before this book was written, I produced a loco design using ready-made Roundhouse components that were available at the time. Dacre featured simple platework that was formed out of standard-width bar and angle and could be easily bolted together. I understand that a set of full-size drawings and a booklet of detailed instructions are still available from Brandbright Ltd. This project became the basis for more ambitious scratch-building Dacres and a good pool of experience has been built up.

The basic principle of building a steam loco around ready-made components remains a 'safe' introduction to the subject. You are virtually guaranteed a model that works, meaning that you can concentrate your efforts on the cosmetic design. So, a little further on, we will look at Mark 2 Dacres, in a variety of guises. But to begin with, let us have a brief overview of how that basic version went together.

THE ORIGINAL DESIGN

Roundhouse cylinders were connected to the rear driving wheels by ready-made slip eccentric gear and rocker arms. The spirit-fired pot boiler was also bought from the same source. An inner wrap-around firebox, the spirit tank and the basic smokebox could be folded up and soldered from this company's etched brass components. The platework was just an exercise in rectangles, mostly of standard widths to save on cutting. The original engine had an open cab and had an early type smokebox door. But after reading the preceding chapters of this book, you may now conclude that the actual outline of a simple engine is a comparatively small matter. Many builders have fitted cabs and more conventional smokebox fronts. A further alternative is to use the dodge mentioned elsewhere, in which the working parts are built in a functional way, then put together as a drop-over tram engine body.

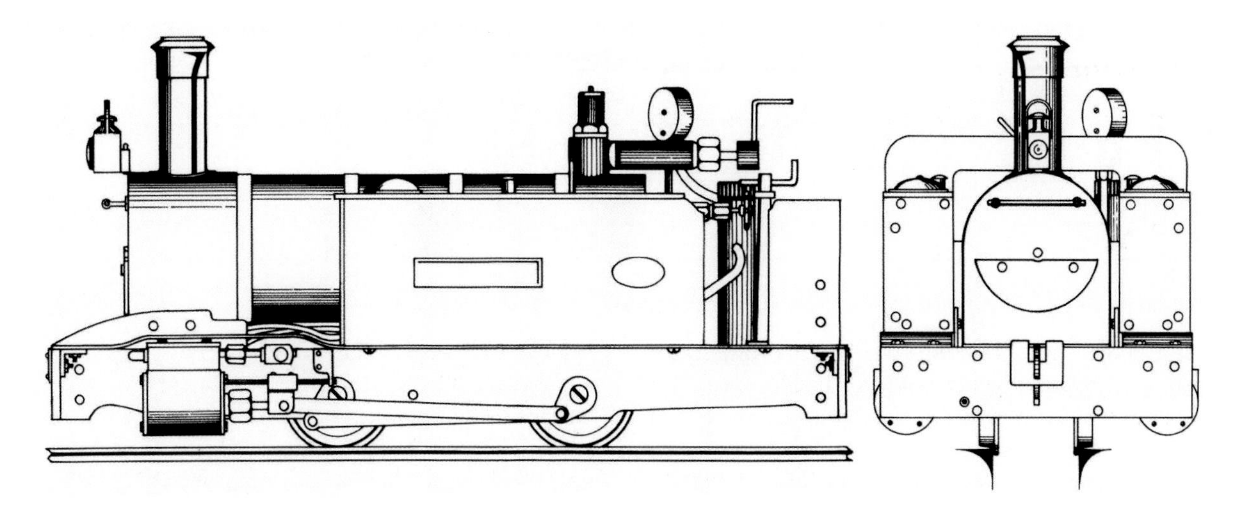

The original Dacre design.

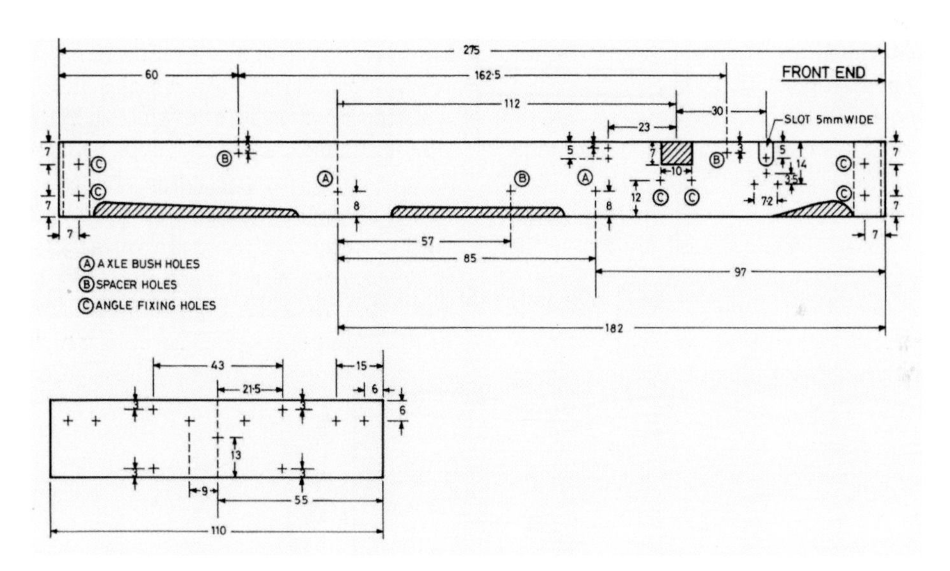

Dacre frames and buffer beams.

I specified thick, rugged material for mainframes, but, even so, it should *just* be possible to open out the gauge to 45mm. A more radical thought for the ambitious would be to put larger diameter wheels outside the frames and build the outline to 2.5in gauge, as a standard gauge tank engine. It would still be easier, for a beginner, than building a full-blown model engineering design.

As can be seen from the drawing, there was one little extra metalworking feature. The running board has a 'joggle' put in it to clear the cylinders. Fortunately, there is a simple and reliable way of achieving this. It is done by wrapping the metal strip in protective masking tape and then squeezing it, in a vice, between strategically placed pieces of metal that have rounded ends. Don't bother to calculate the length of the metal, allowing for what is lost in the joggles. Make the strip too long, put the 'set' in and then cut it to length afterwards. This turns the vice into a rudimentary press tool.

The notion of using a vice as a press tool has many applications. Many a U-shaped smokebox wrapper

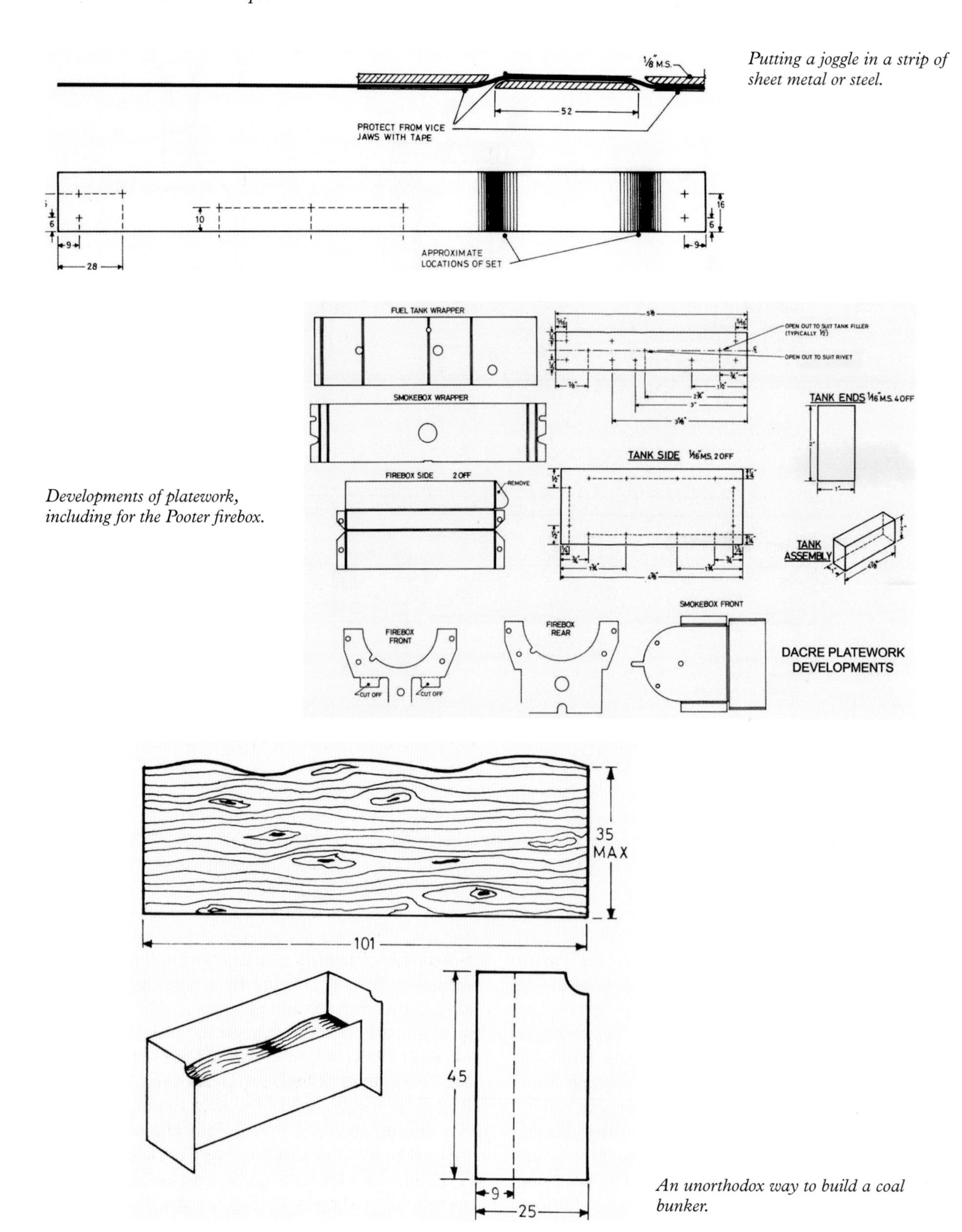

Putting a joggle in a strip of sheet metal or steel.

Developments of platework, including for the Pooter firebox.

An unorthodox way to build a coal bunker.

A cab with a rounded-off roof looks nice on a Dacre.

can be squeezed around a stub of scaffolding tube. As your loco-building adventures continue, you may well build up a little collection of stubs of bar and strip (mandrels) that you can form things around. Always protect your work from the serrated – and usually grubby – vice jaws by tape and/or card packing pieces. You can buy soft jaws for vices, but, somehow or other, I never got round to doing so.

The original Dacre used ready-etched components in thin brass. These are no longer available but they are not difficult to make. For really thin components, there is a way of breaking the rules and using sharp kitchen scissors. The secret is to scribe the outline *deeply* first. Don't exert too much pressure with your scriber initially, as it may want to wander off line. Just build up with repeated gentle strokes until a deep line is visible. Scissor cuts can then follow the scribed line accurately. The edges may want a couple of strokes with the file to neaten them up.

All of the components were assembled using nuts and bolts alone, although the ambitious could use rivets. It is always worthwhile buying proper hex-head screws for the sake of appearance.

To drill a hole for a specific size of screw, you would use a drill that is a 'clearing size' for it. If you are completely new to the subject, there is a good chance that you won't have full ranges of drills. The right thing to do is to buy a specific drill for the job in question. However, if you are stuck, and have a range of old anonymous-looking drills in a tin, you can always try drilling a test hole or two in a piece of scrap material in the hope of finding the right size. Drills have their sizes stamped into them, but the numbers can become indecipherable with age and use. This model, like so many in the hobby, can be held together with 8BA hex-head screws and nuts. The clearing drill for this is 2.2mm (or Number 41).

A lot of construction will consist of screwing various pieces of mild steel and 3/8in brass angle

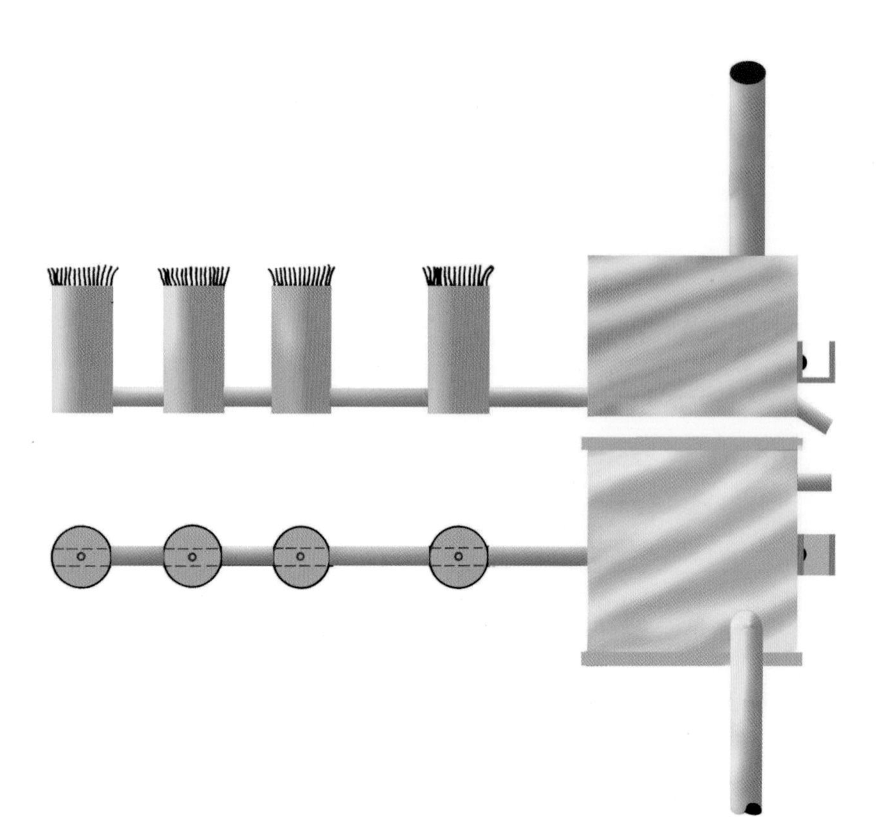

The spirit burner is to a conventional outline.

A rare photograph of the original Dacre on my own railway.

together, as described in Chapter 3. The cylinders are standard Roundhouse and call for no comment. In many engines like this there will be several critical measurements to be made, so that the wheels and cylinder assemblies interact with each other properly. One useful dodge found in Dacre was to build the coal bunker around a block of hardwood – the top of which was wavy – so that, when loose coal was glued on, it looked like a natural pile.

UPDATING THE DACRE PRINCIPLE

The world moved on and Roundhouse moved with it, discontinuing spirit firing in favour of providing simple packages of gas-fired boilers/burners/regulators – complete with a boiler certificate. These represent an instant no-fuss 'drop-in' that makes life very easy. If this is combined with a ready to run chassis, you can see that a large part of building a steam engine has been done for you. What remains can be done with simple hand metalwork, without the need for a machining capability.

The simple Millie chassis and boiler assembly can be a clotheshorse for many a drop-on body. The Glyn Valley Tramway (GVT) used Beyer

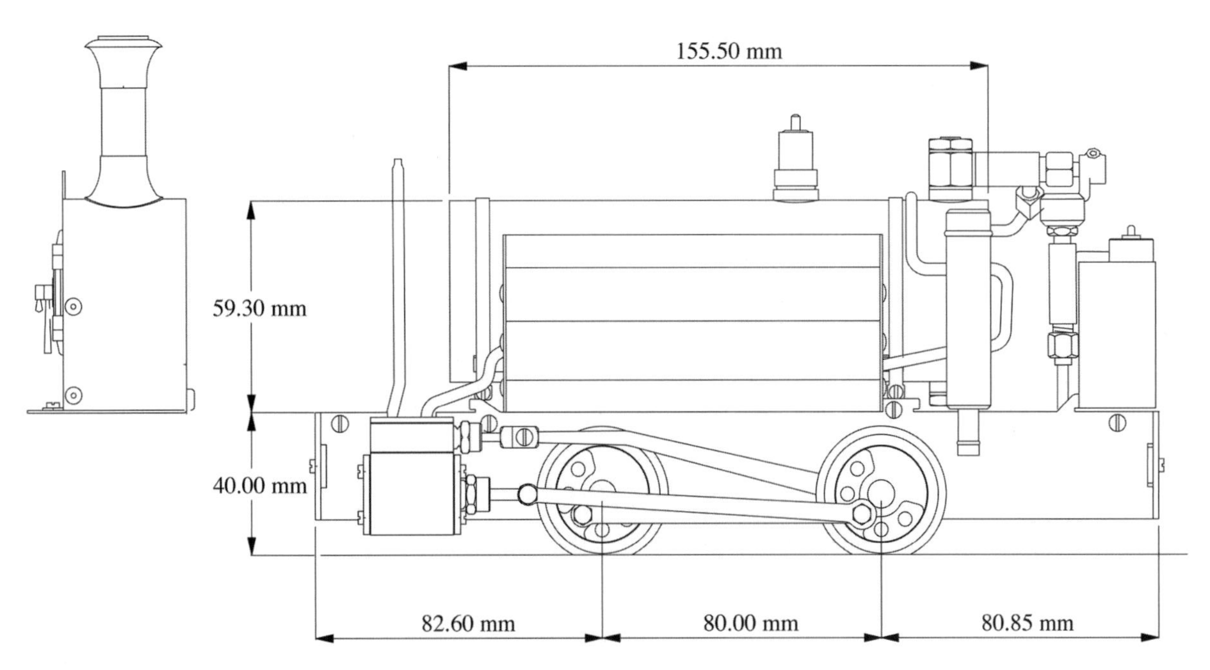

A Millie chassis and boiler. (Drawing: courtesy of Roundhouse Engineering)

Peacock tank engines that would be perfect for the job. It is a representative of the huge variety of standard and narrow gauge tram engines that can be built in the garden gauges. In this case, we are thinking about more substantial engineering than a converted Mamod. Here is a strongly made metal box that conceals a substantial working boiler and chassis, intended to do a good job of work. Basic Roundhouse Engineering components in a raw, functional state could simply be covered with a bodyshell to conceal the non-prototypical mechanics. We are at the next stage beyond a plain outer box – we are starting to see things that are more engine-shaped.

The GVT outline, to my eyes, is particularly attractive. The whole secret of its character lies in that basic square box with rounded corners. It is important to get that absolutely true. I would be looking to buy a piece of brass sheet that would be long enough to form the complete wrapper (for a 16mm loco allow for about 60cm). I would then do my utmost to ensure that it was the right width, with absolutely true straight edges. If I was really new to this, I would saw outside the scribed line and then clamp a sacrificial straight piece of metal in place and file down to that.

Having got that perfect long, thin rectangle, the next thing to do is form it truly. Mark with a pencil where the material is subject to bending – perhaps strips about 20mm wide (it isn't important to be exact). Aim a blowlamp onto these strips and heat each in turn up to bright red heat. This will anneal the metal at just the right areas (that is, soften it for bending).

Here is a good example of where a jig will be useful. Imagine a flat piece of wood with four pieces of wood dowel (of the right radius) sticking upwards where the corners would be. At their tops, another piece of flat wood fits over. What we have are those pieces of dowel truly vertical and well supported – one for each corner. Because strong pressure will be applied, I might just screw this jig down onto the workbench (or hold in a Workmate). Make a pencil mark across the middle of the length of the brass and then offer up the job to one end. Using some flat offcuts of wood, press the brass round at either side. If it is thick brass, it might want to spring back a few degrees from a true right angle. You will

Denys was originally designed to go over a 0-6-0 chassis, but would fit just as well onto a Millie chassis.

need to offer these bends up to an individual stub of dowel held in the vice to force them past the 90 degrees a little so that they spring back to a right angle.

You would now have a U-shaped piece of brass. Clamp a couple of offcuts of wood outside the long sides (to stop the brass springing out) and then force those long sides round the other end of the jig. They will be too long at this stage and will overlap. Don't cut them to length before bending. The chances are that they will be wrong. Only when you have that perfect rectangle plan view (with rounded corners) should you clamp the loose ends together and then saw them both with a single cut. Using a small brass backing plate, solder the two ends together. It may seem a lot of fuss for an apparently simple component, but it is the main component of the bodyshell and, by doing it the right way, we have got it true and accurate, despite any lack of metalworking skills we may have. In short, a splendid example of a jig making accuracy possible. There would be a similar wrapper that forms the buffer beams and side skirts.

The rest of the construction of the bodyshell consists of simple cutting to shape and fixing together. Big areas of brass need a lot of heat and a blowlamp is easier than a soldering iron. Clamp the parts together and apply flux and solder. Each stage will probably require a bit of thought beforehand – but that's part of the fun. Alternatively, you could screw parts together, using pieces of angle and backing plates.

Denys (one of the prototypes was called Dennis), represents a generic GVT engine. You could build a specific example if you wished. If you are looking for a standard gauge engine that looked like a *proper* locomotive, but which had side skirts, I commend a study of the Bideford, Westward Ho! and Appledore railway in North Devon. There were three chunky little Hunslet tank engines, which must have been designed specifically for novice modellers a hundred years later.

A more refined example of the Dacre principle is Java – a design for a generic plantation loco – based around a Roundhouse Sammie chassis and boiler. The drawing of the mechanical parts (reproduced by kind permission of Roundhouse) shows enough hardware for a virtually instant engine. I have merely added a more unusual locomotive design to go around it. The platework is virtually all straight lines and can be built as a separate drop-on unit. Indeed, there are some of us who remember our penurious days when several different bodies were built to go over a working core engine! The only apparently difficult thing to make is the chimney, although close examination shows that the saddle is still in place. In fact, by cutting the top lip off the chimney, it can remain undisturbed and the spark-arresting chimney dropped over. Somewhere in this world, there will be cheap copper ornaments, from which slices can be cut to form the chimney, without the need for any challenging metalwork. The correct name for this is sort of thing is *objets trouvés*, but we know it as having a really good scrapbox.

The domes can be made from plumbing end blanks, which have been suitably annealed and tapped over a bar to form the saddle. Other than that, construction would be very straightforward and the end result would be a colourful and unusual engine similar to this.

A steam street tram and trailer would also suit a Millie chassis as a simple first project.

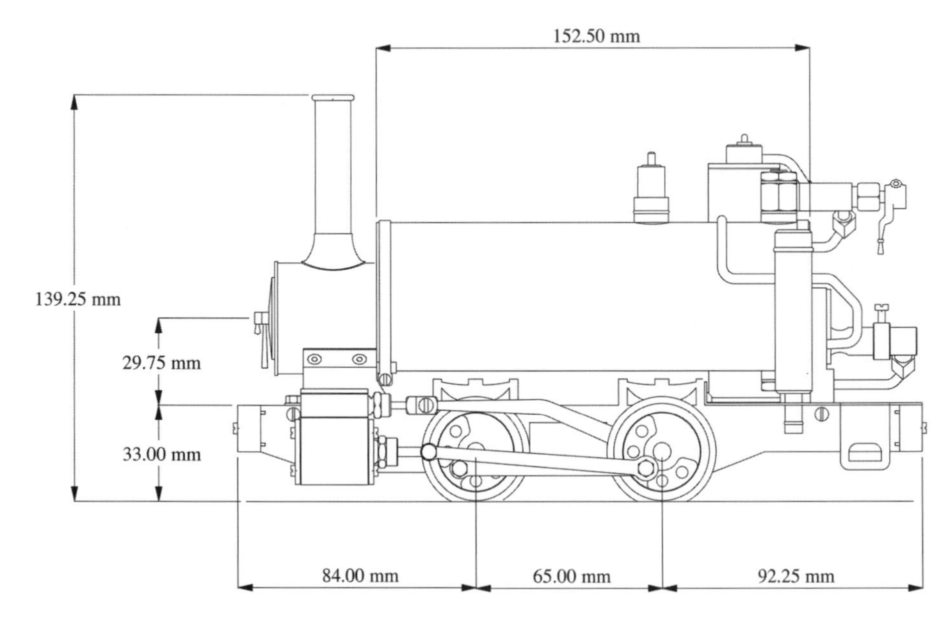

A Sammie chassis and boiler, available off the shelf from Roundhouse . . . (Drawing: courtesy of Roundhouse Engineering)

... this provides a good basis for Java.

Slightly closer to home we have Bardot, a name that resonates with gentlemen of my generation. We have a typical French tank engine of pleasing aspect. The platework is simple in the extreme – just straightforward cutting and bolting of brass sheet and angle. The chimney is unchanged apart from that swivelling cap. It is interesting how a small detail like this can instantly transform an engine.

The domes of this engine were made from plumber's fittings.

Some sort of gloriously French headlight is called for. I have made examples like this in the past by sandwiching copper rod with soldered brass discs and then turning the assembly in the lathe. A vertical hole is drilled in the body of the lamp and a solid section run right through it. But there is surely scope for a rummage through the scrapbox to provide something suitably heroic.

THE DUTCH CONNECTION

For some reason, the original Dacre design was taken forward to form the basis of scratch-built locomotives in Holland. I am indebted to Erik-Jan Stroetinga for kindly providing me with a large file of photographs of the progress made by himself and friends. It is interesting to see how the original Dacre principle was taken on into the realms of pure model engineering. It is with pleasure that I can share some of them in these pages. They are good examples of a way of moving on to

A more refined Roundhouse chassis gives further scope for a simple drop-on body . . .

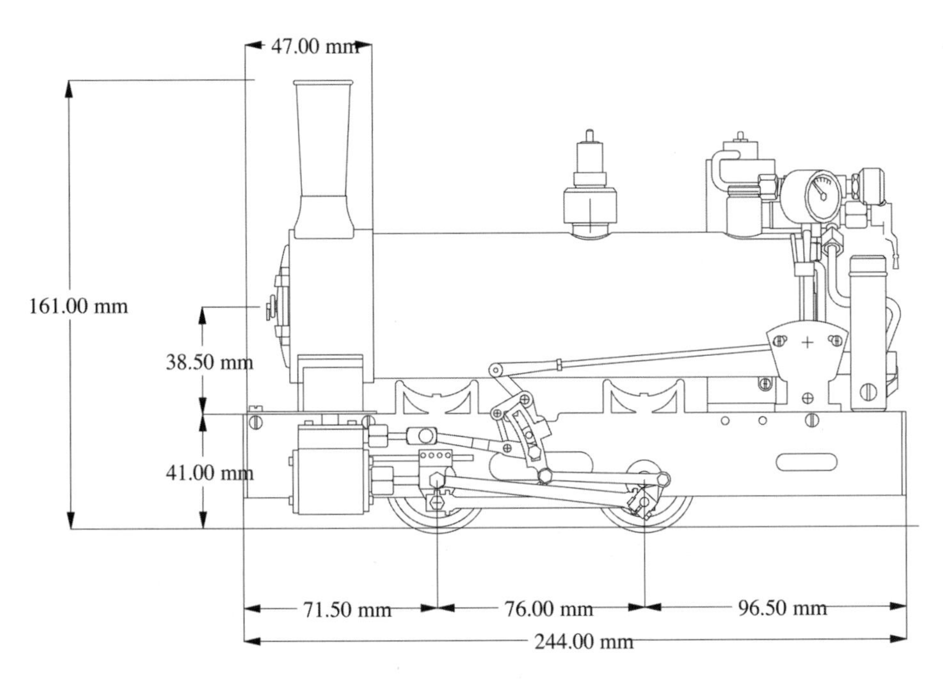

. . . such as Bardot: French to its fingertips.

Roundhouse makes much use of a banjo fitting. A hollow screw goes into a large opening. It is a rugged way of fitting one steam component to a pipe and is easier to make than a gland.

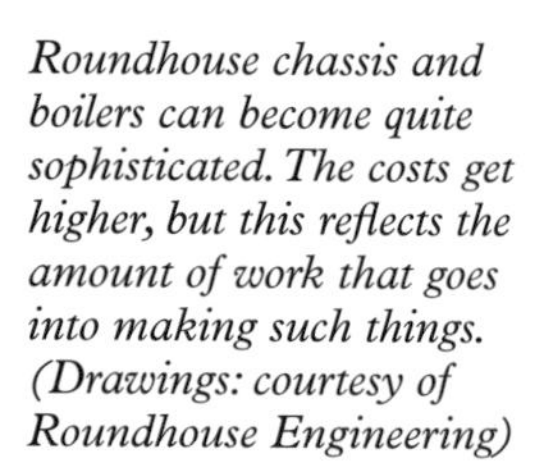

Roundhouse chassis and boilers can become quite sophisticated. The costs get higher, but this reflects the amount of work that goes into making such things. (Drawings: courtesy of Roundhouse Engineering)

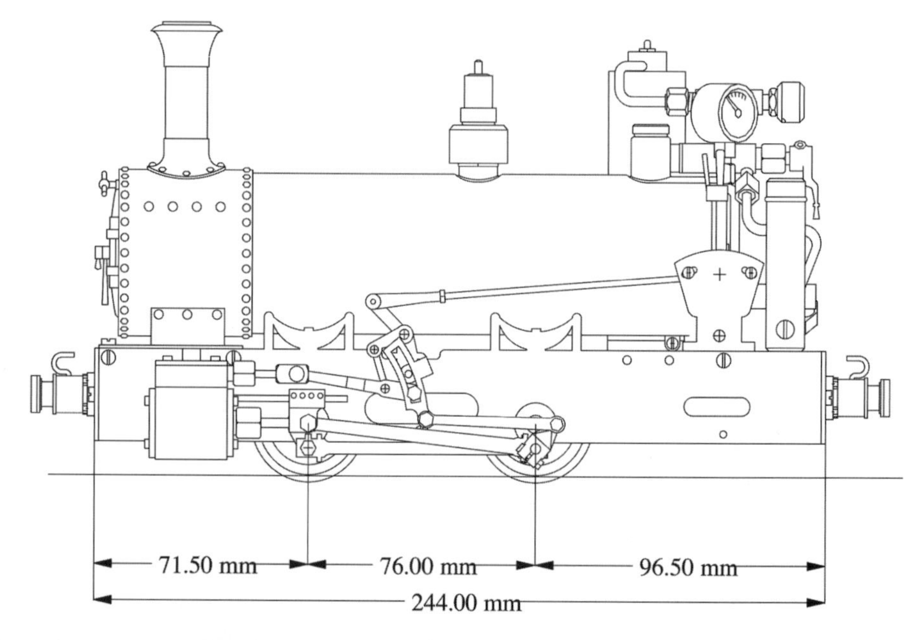

A commercial model of Russell. All of the elements we have discussed so far have come together.

complete scratch-building, for those with such ambitions.

However, a lifetime of loco-building without machining skills and equipment remains an enjoyable option. Although the emphasis has been on Roundhouse chassis and boilers as a basis, there are alternatives. The inexpensive Regner steam engines – such as Willi and Conrad – look a bit shiny and functional, but they are superb slow runners. They too could host a variety of more realistic guises. It is an altogether enjoyable way of producing interesting and unusual working steam engines, limited only by imagination. But to go down the road of scratch-building engines, we need to set about developing some machining skills.

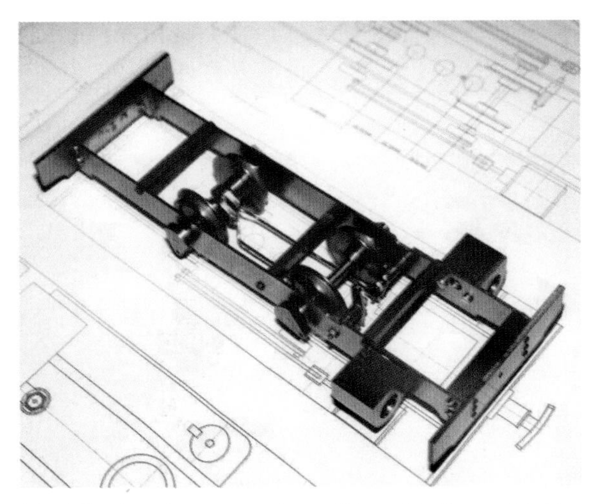

The following photographs have all been kindly provided by Erik-Jan Stroetinga; they start off with a scratch-built Dacre chassis.

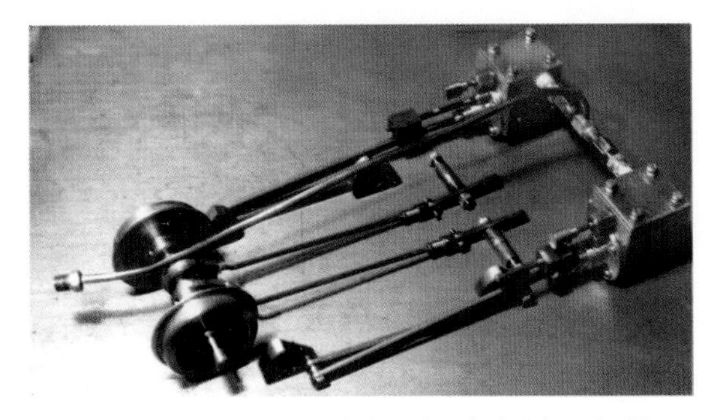

Fabricated cylinders derive their valve timing from inside motion – a refinement of what we saw with Dempsey.

Testing cylinders on air.

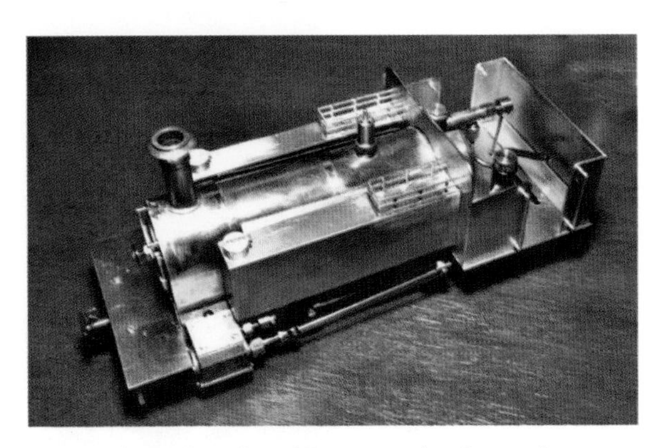

This engineered version of Dacre remains close to the original design, with just a few changes.

Further examples drifted slowly away from the prototype outline.

The gas firing on this example of Dacre is evident.

Putting the gas tank inside a dummy side tank is an elegant solution.

This Dacre backhead has been refined with neat handles.

And finally – a Dacre out in the garden where it truly belongs.

The Dacre principle was used here to produce a Darjeeling engine from the original basic outline. Roundhouse Engineering subsequently developed a commercial model of a Darj prototype.

Case Notes:

Lagos

There is a small but growing interest in ⅞in scale steam locomotives that run on 45mm gauge track. This represents the 2ft gauge (and thereabouts) of the prototype but with added meatiness. As an easy introduction to building a live steam engine in this scale, I offer Lagos. This is a close approximation of a Nigerian steam railbus (tram?) that has been simplified a little and takes into account commercial ready-made components. Roundhouse cylinders and shortened valve gear are suggested. Life is much easier if a gas-fired boiler is used, as there isn't much camouflage for a spirit tank.

When I built something along these lines in the distant past, my starting point was a piece of chunky aluminium sheet that had originally been a large kicking plate on a door. This was cut to form a true flat 'footplate', and was a satisfactory datum level for the entire model. Things could be bolted or riveted to this quite easily.

The lovely wooden hutch would provide cover for radio control and a gas tank. The displacement lubricator would live in one of the side tanks. You will have read enough in this book by now to see how all the elements seem to fall into place. The real engine was a dainty prototype, thus ready-made components for 16mm scale will seem usefully delicate in ⅞in scale. It can sometimes take a mental leap to readjust to this scale at first. To help in this, draw a figure about 130mm tall on a piece of card and then cut it out. Keep 'seven-eighths-scale Bert' handy and he will help you retain a feel for proportions.

And really, that would normally be all the description needed to build the loco. In the normal course of events, you would find a scale drawing of this engine – or some similar prototype that took your fancy – and you could start building. However . . .

Lagos represents the Dacre principle, stretched to a ⅞in scale design.

Lagos *continued*

. . . there is another option here. This design would lend itself to very simple building without a lathe. Let's suppose that, instead of that bogie at the rear, we put in a single wheel set. This could be driven by a couple of reversing oscillating cylinders taken from a Mamod stationary engine (typically an SA3a), via a couple of spur gears to reduce the speed. That lovely wooden hutch would provide cover for any dubious mechanical arrangements. The boiler arrangement remains the same. However, the front pair of wheels, complete with dummy motion, would swivel vaguely in the manner of a single Fairlie on the Ffestiniog Railway. Anything mechanical there is purely cosmetic. I would think that a gas-fired boiler is still called for, but there really is a nice simple steam engine just asking for a beginner to build.

Case Notes:

Scrapbook

In addition to the other designs and studies in this book, I now offer small glimpses into my personal scrapbook in the hope that further inspiration may be found. Some of the drawings are literally hand-drawn doodles and were sometimes penned by a childish hand. But, looking back on them, there is a certain naive charm and they provoked some early models. Some evolved into more serious designs and I suggest that they still raise some interesting issues today.

Greenly

It is this drawing that got me started building steam engines. As a small boy, I was invited to visit an elderly gentleman who had heard of my obsession with trains. In a tiny, cluttered shed, there were bits of small engines scattered everywhere. The highlight was a small tank locomotive that was held in a cradle and steamed up. I was hooked for life – especially as he gave me a general outline drawing, which you see here. Years later, I concluded that this was based on a Henry Greenly design for O gauge (hence the name). All the basic elements of a four-wheeled simple engine were there, including a speed reduction from an oscillating cylinder to the wheels. It was done by a pair of loops of stout thread. We may smile now, but it worked. There have been huge numbers of designs based on this principle. I particularly like LBSC's Gauge 1 version of Juliet, a famous but simple tank engine of much antiquity that has a niche in the history of small steam engines.

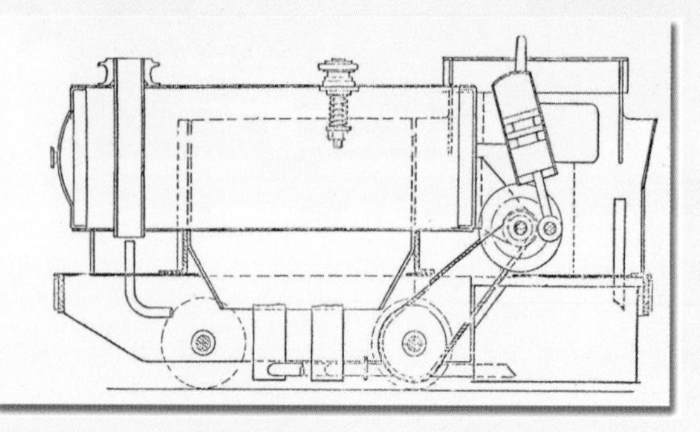

The drawing that started it all for me – Greenly.

Brecon

My early version – *circa* 1952 – of Greenly was as a narrow gauge engine running on Gauge 1 track, so it was quite a large object. The driving cords were replaced by Meccano spur gears. The cylinder and inner boiler were by Mamod, while the bodywork was made mostly of tinplate (cut from biscuit tins) bolted around pieces of Meccano axles. It had no reverse gear and so had to be picked up and turned round at the end of its 9m run along wooden rails. Soldering copper pipe was done with a large dumb iron, heated in the fireplace. It was crude, but it was my first-born and I can still remember with some affection the thought processes that went into building it.

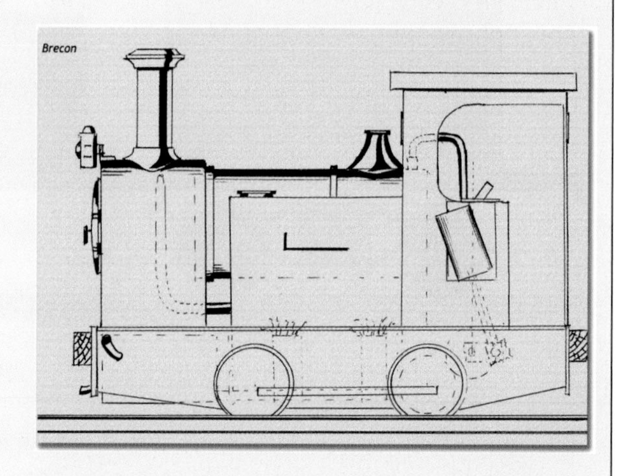

Brecon was an early creation. All the basic elements of a pot boiler are present.

Scrapbook *continued*

Cranefly

A later variant of Brecon; the drawing was still crude but the design was starting to show a hint of modern practice. This was a Gauge 1 standard gauge loco. It was powered by two reversing Mamod cylinders in the smokebox, again with a gear reduction down to the front axle. You will also see a displacement lubricator coyly peeking out from inside the cab. The crane was operated handraulically. As I write these words, I find that there is still a temptation to build a 'proper' steam crane loco – perhaps in 2.5in gauge. There are so many delicious prototypes to choose from.

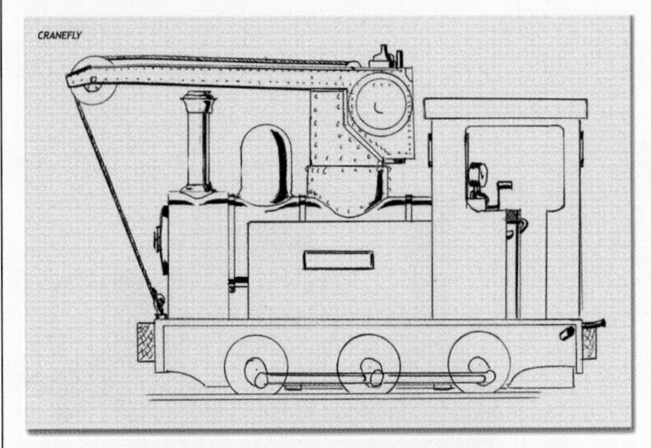

Cranefly still has an appeal all of its own, if you can imagine the completed engine from my childish drawing.

Lewin

This was an early attempt to capture the spirit of a primitive mineral engine. It was going to be a standard gauge engine running on 2.5in gauge track. A start was made but it was beyond me. I could think of no way to hide geared oscillating cylinders at the time. Years later it struck me that I could be radical and put a cylinder inside each side tank, driving a layshaft that ran under the boiler. There would then be a reducing spur gear drive to an axle. But then . . . by the time all that had been done, perhaps one might as well put proper slide-valve cylinders in, to drive a crank axle. And it never did get built. But a coloured enhancement of the original doodle shows that it could still be a very pretty locomotive. I recommend that you try to track down the elusive writings of Brian Clarke. He was the master of such designs and produced some fine examples that registered highly on the quaint scale.

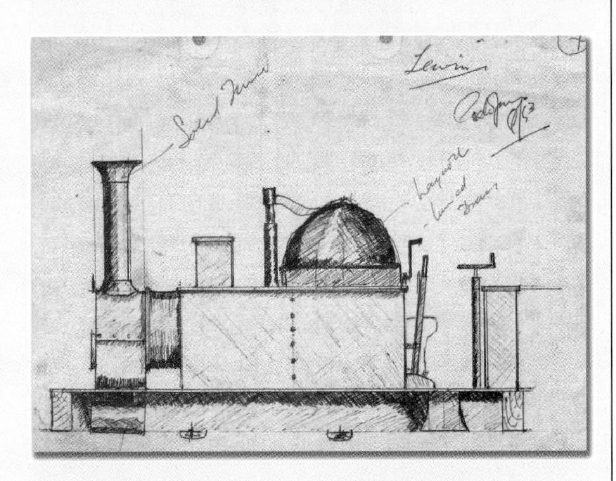

This doodle of Lewin spawned many thoughts. Years later I found an echo of these in Brian Clarke's approach.

An updated drawing of Lewin. The exact shape of the top of the firebox would depend on what sugar shaker top I could find!

Scrapbook *continued*

Gnat

Dating back to 1958, Gnat was a Gauge 1 standard gauge 'low profile' engine, intended to work in a gasworks location with limited headroom. I was still using geared oscillating engines, but this design was getting much closer to a prototype. The footplate in the cab area is located at the bottom of the frames. There was a rudimentary windshield. As I recall, it was not a great steamer – it really needed another wick tube in the smokebox area. The design continued to attract me and culminated, years later, in a 5in gauge passenger-hauling, coal-fired version called Dante. You will see this described in Chapter 8.

Gnat: another very early drawing that led to better things.

La Petite

Here we have a tiny Bagnall export locomotive. That fully enclosed cab offered the opportunity for some very radical thinking. The boiler tube ran right from the back of the coal bunker to the smokebox door, but from the back of the front wing tank it was largely cut away, with just the top curved segment showing. The front of the boiler itself was a disc level with the back of that wing tank. The wick tubes of a spirit burner were located between the frames, stretching to the back of the bunker. The tank for this was located between the frames behind the front buffer beam. A single oscillating cylinder (reversible from a Mamod traction engine) was mounted horizontally underneath the top of the boiler inside that front wing tank. It then drove the front axle via a three-stage reduction gear train, made of Meccano gear wheels. This locomotive ran beautifully slowly. It was eventually sold out of service and I lost track of it. However, maybe a Mark 2 will get built some day and I can enjoy the realistic slow speed once again.

Prince

The primitive plateway systems survived a surprisingly long time in light industrial use in obscure locations. They had a fascination all of their own, with flangeless wheels and L-shaped rails to guide them. Following

The old mock-up of Prince was unearthed for this photograph.

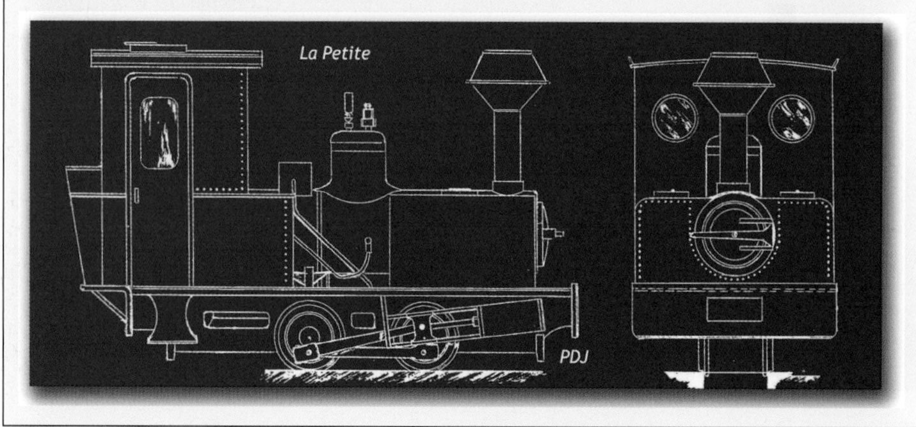

La Petite: opportunities for unorthodox design.

Scrapbook *continued*

Prince.

some experiments, I found that the smoothness of running was quite surprising, even over a very short section of 'cast iron' (moulded plastic!) fishbelly rails. I resolved to build a working steam locomotive one day, but, as a temporary measure, built a simple mock-up, called Prince. You will see from the drawing that there is a central layshaft that connects to the wheels on one side by gears (I have shown the gear on one wheel only). When the elusive 'one day' comes when I build the loco in the equivalent of 2.5in gauge, I will install a further intermediate shaft of speed-reduction gearing. The cylinder and flywheel are on the back side of the engine and not visible in the drawing. For durable outdoor use, I would make the track by soldering lengths of brass angle to the tops of brass plates that represent the tops of the stone blocks. Points would have to be swivelling stub points. Some day my Prince will come.

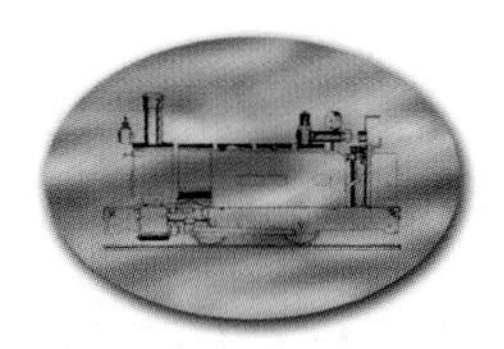

CHAPTER 6

Machining: An Introduction

THE LATHE

Much has been written about the metal-turning lathe and its use. I offer some suggestions for further reading in the Bibliography. The book you are reading now is not a full treatise in workshop machining practice. Instead, it concentrates on how such things apply to the building of our small dragons. The world has moved on since I had a patched-up treadle lathe, upon which I worried bits of brass into the shape of chimneys and buffers. Life is much kinder today. We have a choice of affordable lathes in all sizes and shapes. Carbide-tipped tools are more common. If you buy a new lathe today, it could well have a chuck guard that is interlocked with the motor. And there may well be a big 'chicken switch' for knocking the power off quickly.

Trying to suggest the 'best' lathe for our needs is unhelpful. I built a 3.5in gauge engine on a miniature lathe – a Unimat SL – with many sneaky and dubious practices. It is possible but not recommended. On the other hand, there is no point buying and installing a huge and very expensive lathe in the belief that it will make you a better craftsman. In my experience, it merely shows up any shortcomings on a grander scale. I currently use a lathe that has a 9in swing (4.5in throw) over the bed, and is 20in between centres. That equates to 230mm × 500mm. This is more than enough for small steam engines and would allow me to build locos in 3.5in

The saddle, cross-slide and tool holder.

and 5in gauge. What this extra size does give me is rigidity – a valuable commodity. But this is my personal preference; something smaller is more than adequate.

Don't be tempted to buy a huge battered old lathe cheaply because its sheer size looks impressive. It is possible for someone who knows what they are doing to turn out good work on a badly worn machine, but it really is a case of making life difficult for yourself.

I like a lathe mounted a bit higher than recommended, to avoid the need to stoop as I work. Just above it, I have a couple of flexible spotlights, which give me a good working light. I *try* to keep it clean and tidy, without being obsessional. The swarf is brushed away, while the white tray and splashback are given a quick spray with WD-40 to soften the spatters of gunge before wiping them off.

We know that the lathe's main purpose in life is to turn round things with considerable accuracy. It does so by presenting the tip of a tool to the job at its centre line. The top of that tip is angled back slightly (called top rake), so that the action of the tool is one of planing. With the right tool, right material and the right turning speed, a long continuous sliver of swarf is pared away. In industry, sometimes the turnings are so long that the swarf becomes a nuisance and needs a device to break it up as it emerges, hence 'chipbreaker' tools.

Machining between centres.

The tailstock does more than support the tail end of a job. It can hold drills, taps, dies and reamers.

Speed

That phrase, 'the right turning speed', is an important one. Correct cutting speeds can be found in rather thick books of tables. Different materials like to be cut at different rates. A moment's thought makes it obvious that it isn't just a question of the speed of the turning spindle. What we are concerned with is the rate that the metal is spun past the tool. This will also vary according to the diameter of the job.

I could never remember all the details in those books of tables. I found that, as with drilling in a bench drill, you quickly get a sense of a job turning too fast or too slow. It is less of a problem with modern high-speed cutting tools. In industry, lathes often seem to be hurtling round at impossible speeds. As a rough rule of thumb, it is better to turn a job too slowly than too fast. If it is too fast, things can get noisy, metal starts discolouring and it just seems all wrong. A useful starting point is to set the lathe at its slowest normal speed (without the use of back gear – an additional gear stage that really slows things down) and see how you get on.

With a small lathe, it pays to use the correct coolant/lubricant for a particular material. Large lathes can have recirculating 'suds', constantly playing on the moving job, but they could be considered as overkill for our small engineering. They tend to be used on rougher jobs and, besides, being water soluble, they can invite rust. An occasional squirt of cutting oil onto our job will keep things cool enough. I have a tub of Rocol RTD compound close at hand for all sorts of jobs where metal is being removed or cut into. It goes a long way and my tub will doubtless be used by succeeding generations of the Joneses.

Tool Shapes

Lathe tools come in a variety of shapes for different jobs. Anything with a round nose will produce a smoother cut. Sharp angles will remove metal more quickly but more scratchily. Tools will be straight or cranked, either to left or right, which allows the tip to be pushed into awkward corners. A narrow flat-ended shape gives us a 'parting-off' tool. There are also tools where the tip is at right angles to the body of the tool. This is used for internal turning.

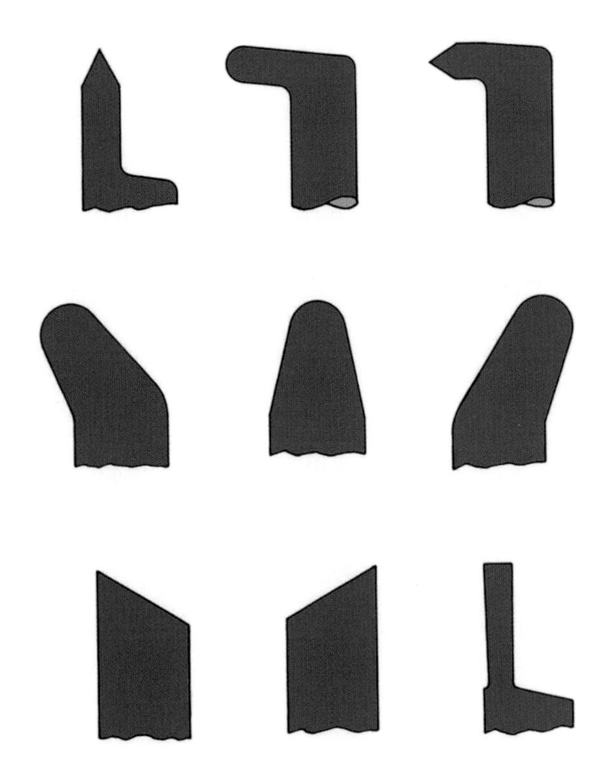

Tool tips show a variety of profiles. Note the round-nosed tools across the middle – good for smooth cutting. The last tool, bottom right, is a parting-off tool.

There is a world of tools that have thread profiling tips, but I suggest that we will do most of our thread cutting with taps and dies. For the sake of completeness, I will mention screw-cutting in the lathe, but only in passing. Many lathes have a longitudinal feed that can be linked to the rotation of the job. This connection is via a selection of changeable gear wheels and means that you can cut a thread in a round bar, using a profile tool.

An archaic swan-neck tool next to a modern carbide-tipped tool.

One tool of considerable antiquity is the swan neck. I like these because they give a really smooth cut. They are not commonly used these days, but if you can get a couple, I recommend them.

CUTTING LUBRICANTS	
Mild steel	Cutting oil (such as Cutmax). For a quick, small job: WD-40
Cast iron	No lubricant used
Brass	No lubricant used
Bronzes/gunmetals etc.	Turned dry, but cutting oil helpful for drilling
Aluminium and alloys	Paraffin
Screw-cutting	Specialized grease

Chucks

The usual basic tool is the self-centring three-jaw chuck. It quickly grips an object to an accuracy of a couple of thousandths of an inch – colloquially known as a 'thou'. This seems astonishingly accurate, but for precise jobs even that is too crude. However, it will come with two sets of jaws for gripping internally or externally and will take care of most of our needs.

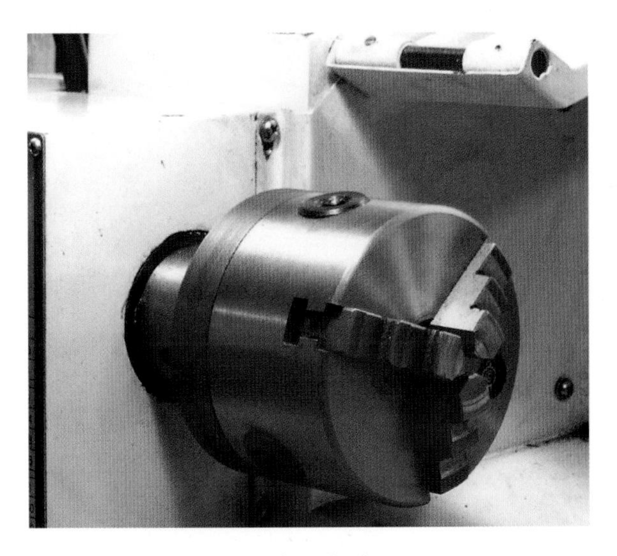

The general-purpose three-jaw chuck.

For perfect accuracy we will use a chuck with four independent jaws (also used for holding non-round shapes). To change chucks, it is a good plan to place something over the lathe bed, just in case a chuck is dropped onto it and causes damage. An offcut of plywood is fine. With the four-jaw, each jaw is adjusted until an item runs absolutely truly. The amount of wobble can be shown with a little plunging indicator bolted onto the lathe somewhere. A pencil held in the tool holder will indicate high spots and thus allow any eccentricities to be patiently adjusted out.

Very large objects can be bolted to a faceplate. Again, we will protect the lathe bed during the changeover. Another thing to be scrupulous about is wiping swarf off any item we are fitting to the spindle. If items bolted to a faceplate have a weight distribution that is substantially off-balance, it pays to bolt a counterbalance weight opposite the heavy portion. This avoids what could be ferocious vibration.

The big faceplate handles large-diameter objects.

Long, thin jobs, such as shafts, can be turned between centres, using a 'carrier' – a device that clamps round the rod and engages with the faceplate. For accurate repetitive work, we might think it worthwhile to buy a set of collets. These are useful if we expect to be making a lot of axles. Flexible jobs may also benefit by having additional support along their length, in the form of fixed or travelling steadies.

Suppose we reversed things and clamped the job onto the cross-slide, or suppose we clamped a short tool projecting from somewhere on a faceplate. This would cut a precise sweeping semicircular recess into a job. This is known as flycutting. Another operation is called boring (as opposed to drilling). If a big chunky cylinder block, which has a rough bore hole already cast in, is mounted on the cross-slide, a smaller diameter bar mounted between centres could be run through it. This bar would have a small cutting tool projecting through it. As the saddle is moved slowly along the bed, the revolving tool slowly cuts an accurate bore.

Boring and Other Tasks

A drill chuck is fitted in the tailstock and it stays still whilst the job held in the chuck revolves. This task is called boring. Given the use of a Slocombe centre drill first, this produces a very accurately drilled hole. If the hole is drilled very slightly undersize, the drill can then be replaced with a correct size reamer – *see* previous Chapter – and the hole will then have a perfectly smooth bore.

However, let us return, for a moment, to the principle of holding a drill in the three-jaw and clamping the job in the cross-slide. If the drill were replaced with an end mill, a true flat surface, slot or groove could be cut into the job as we ran the cross-slide across. The milling process is very valuable to the builder of model steam cylinders, in particular. But there is a problem. During the clamping process, the height of the job has to be packed up exactly. We could do with a device that can adjust the height of a workpiece. Such a device is the vertical slide.

It is a small machine vice with adjustable travel, which is mounted vertically on the cross-slide. So an object gripped in it can be moved relative to a tool held in the three-jaw, both horizontally and vertically. Thus we can mill slots or flat faces onto a job. A job like this likes to have the rigidity of a heavier lathe, but even a moderate size one will let us do the little jobs that are entailed with small-scale steam locomotives.

A Trial Run

Let us have an exploratory go at turning something simple in the lathe. It is easier if it is brass, as this material is more forgiving. Ideally, a short round stub of brass is put in the three-jaw and tightened up hard with the chuck key. *Remove the key!* Yes, it sounds obvious, but most of us forget at some point. When the key flies off it can do all sorts of damage to you, things around you and windows. Make an effort to be obsessed about always removing the key. Pin up a big notice at eye level if you must.

Put a general turning tool in the tool holder, making sure that the tip of the tool is exactly at the centre height of the job. Adjust with packing if necessary (some more expensive lathes will have a device for adjusting tool height). Tighten up. Have a last check that everything is in order – yes, including that chuck key – and then, running at its slowest speed, switch on the lathe. Screwing the cross-slide in, you can feel the tool start to touch the job. Unless it is absolutely truly mounted, the touch will be slightly erratic at first. But as you slowly feed the tool in, the sound of a continuous hiss is heard: a most satisfying noise. Depending on the type of brass, the most likely swarf to fly off will be a drizzle of golden dust.

That was your first cut. Thereafter, you will experiment with different-shaped tools, different metals and alternative cutting speeds. It is a most satisfying learning process. The dial on the cross-slide handle will be graduated in 'thous', or in parts of a millimetre. But if you are turning across the diameter of a piece of work, remember that you are actually cutting away double the indicated amount with each turn (you are cutting away the 'back' of the job as well as the 'front', as it rotates), although if you are turning into the end of a job, this doesn't apply. What you see indicated for each turn of the dial is what comes off the workpiece.

These then are the bare bones of what is involved with lathe work. We will look at tasks specific to building small-scale steam engines in the next chapter.

Vertical Heads

An option with some lathes is a vertical head. This looks like a vertical bench drill that is attached to the lathe. It is used for drilling and milling in particular. If the tool holder is removed from the top-slide and the job is bolted down in its place, it can then be moved in two dimensions on a flat plane. By bringing down the head, with an end mill in place, the job can then be end-milled to a new flat level or grooves can be cut truly in it. It is a particular help in making cylinders. But you might well conclude that it is rather a luxury at this early stage, just for the sake of a couple of tasks. There are usually ways of getting round the problem. Perhaps I could suggest that one of these might be to ask someone else to do that job for you as a favour. Mutual help is a fine tradition in the world of model engineering.

> One option for building a steam engine without any substantial equipment is to consider registering for part-time classes at a college. As well as a magnificently equipped workshop, you will get tuition as well. Some tutors may not know what is involved in building a steam engine for the garden gauges, but are usually amenable to accommodating those needs when you explain them.

The next stage, beyond a milling head, is a dedicated milling machine. For small tasks, such as we might meet when building an engine to the Dacre principle, one of these would be a case of overkill. Not only is there the cost, but they also occupy more space in the workshop. If you have a small travelling machine vice that can bolt to the table of a bench drill, this will cope with all of your needs.

GRINDING

The grinding of tools to sharpen them is a skill well worth learning. A modestly priced, double-ended

A cheap travelling vice, sitting on the table of a bench drill, will allow small milling jobs to be done.

bench grinder will repay its initial modest cost many times over. Before we start, however, I want to reinforce the safety aspects. Safety goggles *must* be worn. The matter is not for discussion. Safety guards *must* be left in place. Grinding wheels should be correctly attached, together with the correct washers. The tool rest should always be adjusted so that it is as close to the grinding wheel as possible, without actually touching.

Holding a tool against a grinding wheel produces a spectacular shower of sparks. These are generally harmless from a heat point of view, but common sense dictates that you should not have any highly flammable objects nearby. Keeping the tool held at the correct angle means that a freshly cut face is soon produced. Most of the skill of grinding is tied up with that phrase 'correct angle'. A simple example would be the sharpening of a cold chisel. You will have noted the steep angle of the cutting edge before grinding. You may well be horrified at how battered and damaged it is; the aim, therefore, is to reproduce that angle in clean, unspoilt faces. There is a knack in drawing the chisel slowly from side to side whilst the grinding goes on, to try to get a straight new face right across its width. The worst of the mangled tip may be taken off with the coarse-grained grinding wheel. When both sides of the tip are done, the job is repeated with the finer-grained wheel.

For a cold chisel, I wouldn't hone the blade to a razor-sharp edge with an oilstone. Such a tool is a blunt instrument of violence. But I might case-harden it. The other thing that happens to a cold chisel is that the end being hit all the time 'mushrooms' out and can be dangerous to handle; a small slip and that mangled end can tear into your hand.

Grinding a big mangled mushroom away can be slow and laborious for a modest angle grinder. My own preference is simply to cut the splayed top off a cold chisel with a hacksaw, then bevel the end inwards slightly on the grinder. My chisels are now 50mm shorter than when I bought them forty years ago, but they will see my lifetime out quite happily.

Lathe tools are quick to sharpen because there are only small amounts of metal to be removed. Examining the tip of a worn lathe tool reveals that the really sharp angle between the top and front

Case Notes:

Anton

This design was aimed at providing live steam power that is LGB (G scale) compatible. In effect, it is a refinement of Dacre. The main feature is that it will go round very sharp curves. You will note the short driving wheelbase and the arrangement at the rear. It is, in effect, a very long pony truck. This calls for a substantial rigid floor around the cab and bunker area that will avoid flexing. However, there is an alternative arrangement that better reflects the prototype. The bunker is attached to the pony itself and swivels with it. This involves extra work in building the model. Either method will allow this powerful steam engine to run on sharp Setrack curves.

You will note that the engine was a conventional spirit-fired potboiler as originally designed, but a gas-fired boiler is equally applicable. The cylinders shown are Roundhouse, with the valves driven by shortened rocker gear from eccentrics between the frames, although even a foreshortened Millie arrangement will work well. In the interest of being a simple engine for beginners, I would suggest avoiding fitting Walschaerts gear in such a short space.

Those long side tanks would be a good warm place to hide a decent-size whistle, if you must have one. Most of the platework is straightforward and flat. This is very much an exercise in cutting shapes out of brass and joining them with pieces of brass angle and rivets. The spark-arresting chimney can be turned from solid, but a beginner should be able to fake up the bulbous portion to drop over an existing conventional chimney. Or, if it is a really simple life you seek, put an ordinary chimney on instead. The engine will still look quite handsome. For that flat-topped dome, a plumber's blanking cap or similar should make life simple (*see* page 89).

CASE HARDENING

It is easy to put a tough skin on the outside of a piece of mild steel. This is sometimes used when making high-wear components in valve gear, for example. The job is heated up on the hearth until it glows red hot. It is picked up with tongs or forceps and the area to be treated is dipped in case-hardening powder (Kasenit being a common trade name). This adds carbon to the outside of the steel and makes it tougher. For *very* precise work, it is worth noting that this layer has a very small thickness of its own, which has to be taken into account. But for something like a cold chisel it is unimportant.

facets has been worn. To get the top rake ground, turn the tool sideways with the top away from you. Offer it to the wheel and a brief touch will produce a nice, shiny finish. Then turn the tool the right way up and introduce the front of the tool to the stone. A couple of seconds of grinding produces a freshly sharpened lathe tool. If you find it to be slightly 'scratchy' in use, reinforce those angles you have just ground by manually rubbing the tool across an oilstone to which a couple of drops of oil have been added.

The big danger with grinding is that too much heat builds up, which can draw out the temper of the steel. Only brief touches on the wheel are needed. If a lot of metal is to be removed, have a small container of water handy in which to dunk the piece regularly between brief grindings. A refinement of this is to have a grinding wheel which sits in its own water bath.

So far, we have been talking about tools made of tool steel/high-speed steel (HSS). Carbide-tipped tools are much more durable and can be used at a higher speed, but when they need sharpening, an ordinary grinding wheel won't touch them; the wheel wears away. Instead, a green grit grinding wheel is needed.

Sharpening drill bits is a much more precise game. You have to judge several angles at the same time. Both sides of a tip need to be ground equally, so that it looks right when viewed end on. You can buy drilling jigs that will do this accurately and I would urge the purchase of one. If a drill is ground unevenly, it will probably cut a hole bigger than the drill size indicates. For non-critical drilling, you should be able to learn to grind a drill, by eye, within a few weeks.

A large, sharp drill bit, rotating at the slowest speed of your bench drill, is a joy to watch. In mild steel the swarf turns away effortlessly, with just a drop of lubricant to keep things cool. If the drill is blunt, it seems to take much more effort and the swarf turns a blue colour due to the heat.

For sharpening wood chisels and plane blades, the grinding wheel is best used for rough shaping only. Most of the work should be done on an oilstone with a roller guide to ensure accuracy.

STEEL SWARF

If you produce a lot of small steel swarf, retain a quantity of it to use as a wagon load for a small-scale railway. It looks quite prototypical when new, but equally interesting when it turns rusty. Such things could be found behind an engine shed somewhere.

There are many things that can be sharpened on a grinding wheel that go beyond the scope of a book on locomotive building, but the usual principles apply to everything. Make sure that the angles are right and don't overheat the metal. Grinding is a specialized subject. Indeed, it can be a lifetime's profession. Our little engines will only call for a moderate amount of knowledge, but the reading of a specialized book on the subject is highly recommended.

These, then, are the basic outlines – starting points for further study. In the next chapter, we will look at machining jobs that are specific to making small steam engines. As well as the technical details, there is a surprising amount of scope for ingenuity in tackling particular jobs. If you can find someone locally who does small metal turning for a hobby – it needn't be model railway engines – try to arrange a visit to get a feel for things. There's nothing like watching machining being done to help you on your way.

Case Notes:

Lyonesse

'Ah, you should go to Ashton and have a peep at Boulton's siding – it is always crammed with historic interest. You'll see things there that are unparalleled elsewhere.' So runs a quotation in *The Chronicles of Boulton's Siding* by Alfred Rosling Bennett. This book, written in 1926, was a reprint of articles that appeared in *Locomotive* between 1920 and 1925. Isaac Watt Boulton ran a small locomotive building and repair business. However, that is to underestimate the sheer variety of small, unusual engines that came through his works and often emerged rebuilt out of recognition. There were many quaint little contractors' engines in particular, often of unorthodox appearance. For the student of unusual locomotive practice, here is treasure indeed.

Lyonesse is my freelance offering that encapsulates the spirit of Boulton's siding. Designed as a small standard gauge loco in 2.5in gauge, it is a geared model with a single cylinder driving a layshaft. A small gearing-down effect comes via the transmission by a Mamod spring belt, but a train of several spur gears would possibly be a better option. The translucent drawing shows most of the main components quite clearly. This is quite a small loco, so a boiler from a Mamod stationary engine might suffice. In the tubular firebox there are a couple of wicks to give most of the heat, although there is an additional burner located in the smokebox.

The cylinder shown is a Roundhouse one but almost anything of similar size will do. A flywheel will be needed on the layshaft, on the opposite side to the cylinder. This would just help to smooth the rotation of the layshaft. But maybe there is an optional refinement here. If the pulley or gearwheel on the shaft has an adjustable grub screw, it could be loosened so that the flywheel would spin but the engine would remain stationary. Thus the engine can be run on tickover in the workshop or on a siding, just for the pleasure of watching it. Suppose we added a pulley wheel outside the flywheel . . . this engine then could drive stationary machinery via a belt. And if you think this is fanciful, read those *Chronicles*.

The boiler pressure could be raised to 20psi with advantage. However, I must add a warning here. Mamod boilers are designed to work at low pressure,

Lyonesse *continued*

with a huge safety margin. If you increase the working pressure, you do so entirely at your own risk. Many boilers have been altered in this way, but I have to make a disclaimer here for this practice. Certainly, tampering with boiler pressure should not be undertaken by children or complete beginners. However, Lyonesse will steam well enough at normal Mamod pressure.

The domed top of the firebox was originally made from a slice taken from a copper ornamental jug. I keep a box of odd little shapes like this – picked up cheaply at jumble sales and the like. If I am stuck for some ornate shape, I can usually find something to adapt. If you can find something resembling the urn-like fireboxes of Gooch's early Great Western engines – a fascinating subject to research – so much the better.

There was a whole world of these tiny, somewhat primitive, mineral engines hauling rakes of wooden wagons. It is my fond hope that many will come to fruition, one day, in the form of small-scale live steam locomotives.

My original mock-up for Lyonesse in 2.5in gauge. We see Bert trying it for size.

CHAPTER 7

Specific Machining

Having had an overview of machining practice in the previous chapter, we can now look at specific procedures applicable to building small steam locomotives. I am assuming that you have a lathe and have some experience in using it, or are in the process of gaining it. If you are a skilled machinist then you have a headstart anyway.

MACHINING WHEEL CASTINGS

Wheel castings usually come in the form of uninspiring lumps of grey cast iron. To run true, they need to be machined very precisely. If you are fluent in the use of a lathe, you will probably mount them in a four-jaw chuck, which you will adjust, with a dial indicator, until they are absolutely spot on. But fortunately there is a less demanding technique which guarantees accuracy for the beginner.

Start by checking over the back of the casting for any undue humps and bumps and file them off (I briefly touch the casting on a belt sander – but only because I have one). Put the casting, front side out, into the three-jaw chuck. Face the rim off with just enough passes of a chunky tipped tool, to reveal the shiny metal underneath. Note the setting of the final cut and apply it as you repeat this job with the rest of the wheel castings in turn. At this stage, the wheel bosses can also be faced off. Check with drawings to note that these usually stand out slightly proud of the faces of the rims. Then turn the castings around so that the shiny faces are bedded in the three-jaw and face off the backs to a common size.

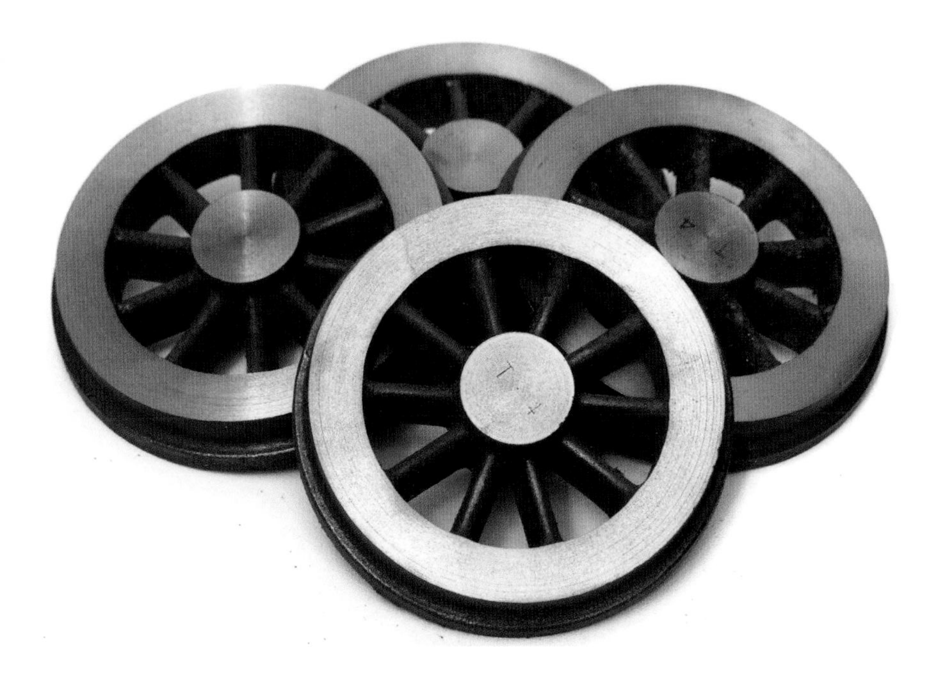

SOFTENING 'CHILLED' CASTINGS

The skin of a casting is usually tougher than the inside, so it is a good idea to try to get under it quite quickly. There is just a chance that you may come across a 'chilled' casting, although they seem much rarer than in years past. You will recognize one of these because it is so hard that a steel tool just seems to skate over it. If you are unlucky enough to get one, put it in the kitchen oven for a couple of hours and then keep it on a radiator for eight hours. This is usually enough to make the internal structure of the metal better behaved. An alternative is to put it in a hot solid fuel burner, fireplace or range and leave it to cool in the ashes overnight.

With the front and the back of the castings roughed out, the wheels can now be faced down to final thickness. Put the first one, front inwards, in the three-jaw and make a few passes across with a round-nosed tool. Make the final cut fairly light and move the tool slowly. This will give a nice clean finish. Make a note of the setting of that final cut and repeat this with the other wheels. When this is done, put a wheel in the chuck, face out. Then turn down the face of the rim until the wheel is to the thickness shown in the drawing. Make a note of the setting and turn the other wheels down to match. If you are unsure of your ability accurately to turn to an identical setting on all castings, then turn that first wheel down slightly oversize and do the same with the other castings. Then make a final cut, with the carriage locked in position. When you make the passes over the other faces, with the setting locked in one dimension, the rest of the wheels have to be all of identical thickness. They can't be otherwise. Repeat this exercise with the wheel bosses.

All of this seems painfully long-winded and slow. No matter; it soon becomes an automatic process and hence much quicker. The process of machining multiple components, to a common setting, is at the heart of much accurate machining.

Drilling Out

Now for a critical bit. The hole drilled in the centre of the wheel for the axle has to be dead centre and at a true right angle to the casting. As the casting is still a rough lump around its edges, this calls for a certain amount of judgement. Using odd-leg (or hermaphrodite) calipers, you can scribe intersecting lines from the edge of the wheel boss. I particularly like to use the inside edge of the rim as a reference point. Centre pop accurately. If you are not satisfied that the pop is perfectly located, you can drift it sideways slightly by tapping it again with the centre pop held at an angle.

The most accurate way to drill into a tiny pop mark is to use a Slocombe (or centre-) drill, held in the tailstock drill holder. This deepens the initial hole into something large enough for a conventional drill to enter accurately and not wobble about. Drill out fractionally undersize and then finish by putting a correct-size reamer through. This last tool should be regarded as a finishing tool, not as a glorified slightly bigger drill. Its sole purpose is to produce a perfect finish to a bore.

If you don't possess a tailstock drill holder, you can drill out (again with a Slocombe drill first) in a cheap bench drill. If you don't possess the correct reamer yet, put the final drill through a few times at high speed. Given care, this will give a true enough hole for our purposes, as we shall see in a minute.

OPTICAL CENTRE POP

As well as the humble plain centre pop, there is a very useful refinement. The optical centre pop frequently consists of a moulded or turned carrier with two holes drilled through. In one of these is a piece of ingeniously shaped clear plastic rod. When you peer down through this, your finely scribed lines seem huge. There is a little spot and a ring marked in so that you can align this plastic very accurately with a specific point on the job. The underside of the carrier has a non-slip surface. When you have lined up the point exactly, holding the carrier in place, pull out the plastic tube and replace it with a short stubby metal punch located in the hole next door. Give it a tap with a light hammer and the pop mark is exactly where it should be. Highly recommended as one of life's little luxuries.

A Mandrel

To get that perfection of turning we need, given limited equipment, we will make a mandrel. The

When you peer down the optic of a centre punch, you see intersecting marks as big as tram lines.

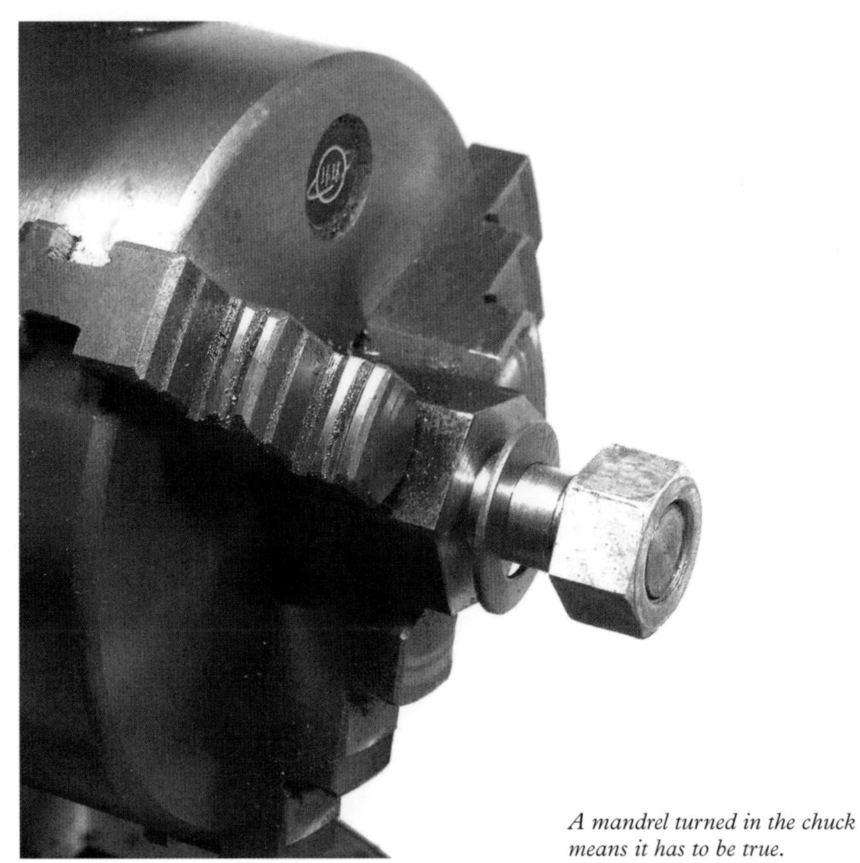

A mandrel turned in the chuck means it has to be true.

first step is to make a centre-pop mark on the three-jaw next to one of the openings for the jaws. Next, we shall need a cob end of round steel bar (though I prefer hexagon bar) off a size appropriate to the scale we are working in. Face off both ends. Put a centre-pop mark somewhere around the curve of the bar (or one of the hex faces) and insert the job into the chuck, so that the pops line up. That way, the job can always be replaced in the chuck, to the same alignment, every time.

Turn down part of the protruding portion of the bar until it will just go – snugly – through a hole drilled in the centre of one of the wheel castings. You now see the reason why we have used a mandrel. Even if your three-jaw is a bit untrue, because the mandrel always goes back, pop-to-pop, the bit we have just turned will always rotate truly. By sliding a wheel over it, this means that the rim of the wheel we turn will always be true to its centre hole.

Something is required to hold the wheel in place. You will need to take a die and cut a thread. If you have a tailstock die-holder, use it. But, if not, you can bring your tool post towards the job and use it as a true flat surface to hold a conventional diestock square, as you apply it to the job. Pull the lathe over by hand and cut the thread. Then it is just a case of adding a nut and a washer.

Turning Rims Truly

Put the first wheel in the mandrel and turn up the diameter of the main part of the rim, with a slight angle, as shown in the drawing. A wheel runs best on rail if there is a slight taper on the tread (set the topslide at the angle required) and a curved root into the flange. The drawing of your particular prototype will show this. Again, after turning down to a slightly oversize setting, do the same with the other wheels. Turn the flange to profile, noting the settings.

Then make a final cut on the tread of the first wheel, with a slightly round-nosed tool. Make a note of the setting and then apply that to the other wheels, so as to ensure all diameters are exactly the same. Likewise with that flange profile. I just round off the sharp corners of the tread and flange

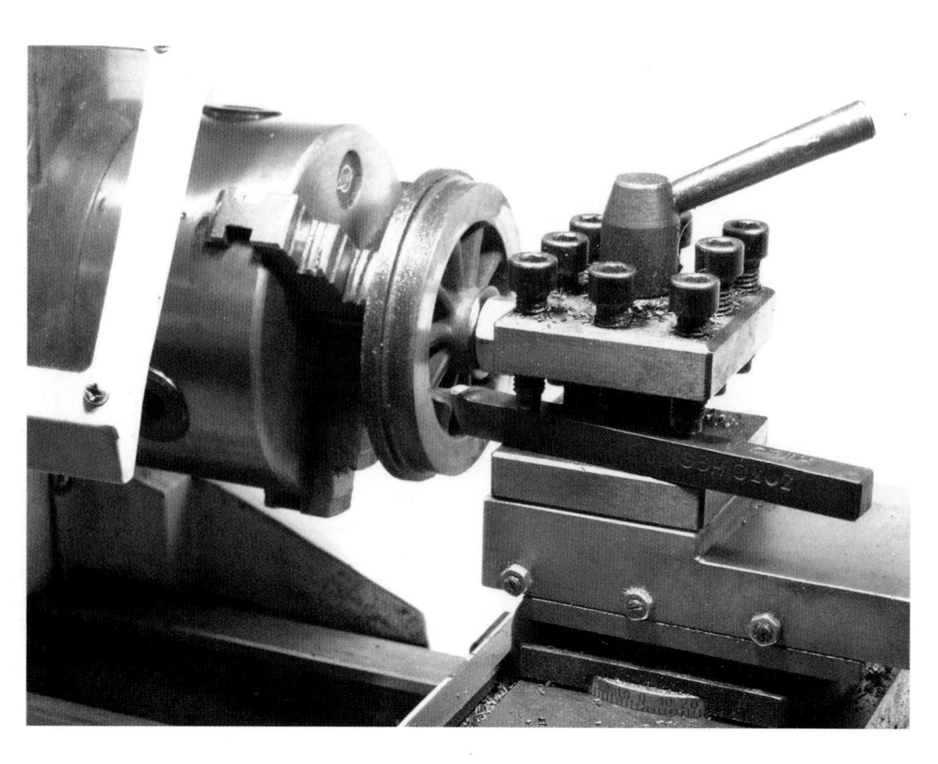

The inside of the rim being trued up in the mandrel. All the driving wheels would be treated to this process, without disturbing the setting of the cross-slide.

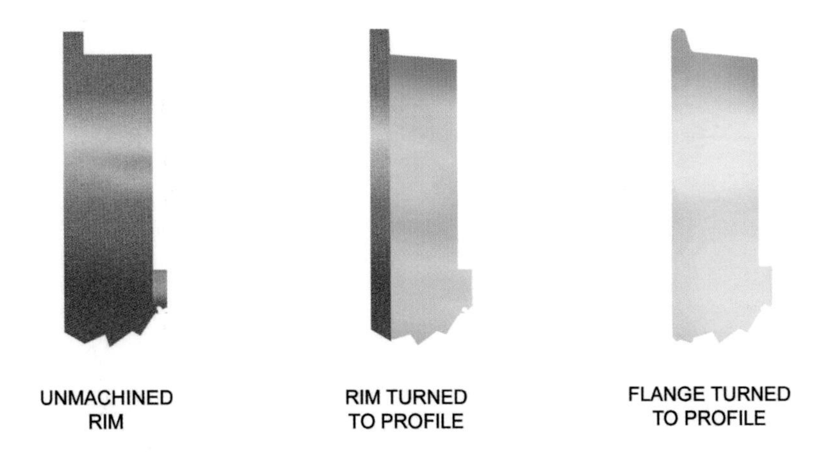

Stages in turning wheels to profile.

with a file, lightly resting on the job as it spins in the lathe.

It won't matter if the wheels are very slightly under or over diameter. In real life, this could vary depending upon how much wear and re-profiling they had received. But you will know that you have done the essentials right – the treads of the wheels are all the same and are perfect in relation to the centre hole. The flanges will all be the same profile. If this was your first attempt it could well have taken a long time, but you will have achieved a high standard of accuracy.

It is good housekeeping now to clean the mandrel and spray it with WD-40, then put it in a plastic bag. Tie a label to it and keep it in a box marked 'Jigs'. You may need it again in several years time, by which time you may have a collection of such things.

Drilling for Crankpins

There will be one more job to do to driving wheels and it may be convenient to describe it here. They have to be drilled precisely for the crankpins. Fortunately, there is a useful dodge to ensure accuracy. First of all, decide what the 'throw' of the pins will be. If it is an outside-cylindered engine, the distance between the centre of the wheel and the centre of the crankpin *must* be half the distance of travel of the piston, taking into account the thickness of the piston itself and any slight recessing of the cylinder covers into the bore. Apologies if that sounds complicated, but a couple of minutes with a sketch on the back of an old envelope should make it clear. If you are working to an established design, the throw will be marked on the drawing of the wheel anyway.

If the loco has inside cylinders then the position of the crankpin merely has to be in the centre of the small diameter of the extended crank. But it is *vital* that the throw of the crankpin is identical on all wheels.

My solution is another little jig. I leave the axles, when turned, slightly overlength, temporarily. The device is just a short length of flat steel bar with a hole drilled near one end, such that it is a tight, but movable, fit over the axle. At the correct distance along the bar a second small hole is drilled; this is deeply countersunk with a much larger drill. The aim is to be able to *just* get the tip of a scriber to protrude beneath, whilst being held snugly. The photograph on page 115 shows this clearly. Then simply scribe a short arc on each of the wheel bosses in turn. So now we know that the throw of the crankpins will be identical.

Scribe a short centre line on the bosses, to intersect the arcs. This can be done by cross-measuring with dividers, but, if fortune is with you, there will be a wheel spoke that is aligned with the centre of the boss. Lay the wheel down on a flat

Case Notes:

Klaus

This is based on a comparatively new build in the Winterthur works in Switzerland. It was designed around very modern technology and examples created quite a stir when they first appeared. To my eyes, they appear rather wide and ungainly, but they have proved to be excellent servants, with many innovative design features. I was indebted to Klaus Matzka for supplying the general arrangement from the works drawings at the time – hence the name in recognition of his kindness.

As a glance at the drawing will quickly reveal, we have a two-cylinder engine here that drives an intermediate shaft. This is then geared to drive a sprocket on one of the main axles. If you can make this a three-stage series of spur gears, you will get tremendous power at very slow speed. As the engine forges up your particular mountain, the exhaust is sounding like an express train and the motion is flying round in a blur, whilst the engine itself steadily forges ahead. This is how all good rack engines should behave, of course.

I suggest keeping the final sprocket rigidly attached, along with the wheels, to its axles. This way, when it is coming back downhill you can control the speed, or even stop the engine, with counter-pressure in the cylinders. On this subject, were I to build a radio-controlled rack locomotive, I would allocate a channel to applying a clamp brake on the intermediate shaft.

Spirit firing is not really an option here, so a gas-fired boiler it has to be, with a flue running down the centre. Given a choice, it would be better if the flue ran below the centre line, so that more would remain covered with water when the engine was down at a level base station, with the boiler tilting downwards in its 'kneeling cow' stance.

Given a bit of ingenuity, one could mount both cylinders, the valve gears and the intermediate shaft on a piece of flat plate bolted to the tops of the frames, so that the entire mechanical gubbins could be installed as a pre-built unit.

For a good part of my lifetime, I failed to think up a simple rack system that would be completely reliable and strong enough to be durable in all weathers except icing. It was only as I was starting on this book that the thought of bicycle chain came along. And in principle it works well. A small tensioning sprocket is fine for use as the final drive from the loco. I had a few experiments and found that by drilling out an occasional rivet and

Klaus *continued*

replacing it with a staple-shaped piece of brass rod, it could be pinned down to the substructure, between the rails. I also found that by removing rivets and replacing them with smaller diameter nuts and bolts, there was enough flexibility to persuade the chain to move in very gentle curves and the sprocket seemed to follow it successfully. Brief experiments appeared promising.

Very large radius stub points would be possible, but, as you gather, much of this is new territory. Conventional points on the flat, rackless system are not a problem, but there is a need to ensure that the outer diameter of the drive sprocket just clears the running rails. There is considerable scope for further experiment. However, there is a lower-tech alternative. The adhesion of model locomotives on smooth rail is proportionally greater than that of the prototypes. Provided that the incline were not too severe, there is a case for building Klaus or similar, as a live steam loco, with substantial gear reduction, but relying on the wheels for adhesion. Yes: plenty of scope for experimenting. Needless to say, the outline of this design is not set in stone. It could equally be adapted around drawings of a Snowdon Mountain loco.

A simple jig for the crank throw ensures accuracy.

white surface and then, with a ruler, draw a pencil line above and below the wheel. When you are sure this line is to your satisfaction, scribe a short vertical line to intersect the arc. It is the throw of the crankpin that has to be precise. Its position along the centre line of the boss only needs to satisfy the eye. But if you feel in need of more precision, use the dividers to scribe further intersecting lines on the wheel, from various points. Hazy memories of school geometry lessons will provide one of many different ways this can be done. Accurately centre-pop the point where the arc crosses the centre line. Use a small Slocombe drill to start things off and then drill through to size. An alternative is to drill the jig out to the required size of the hole and then drill straight through it into the crank – taking care that you are drilling along its centre line.

Again, spray, label and bag up the little jig; another one for the box.

MAKING BRAKE SHOES

There is a nice, simple way to make brake shoes on the lathe, which is also a useful early exercise in turning. A prudent loco-builder will steadily amass a collection of hexagonal nuts and shapes. The most useful are large-diameter thin nuts (I still call them 'gas nuts'), but hex brass fittings can be used for this purpose. Search the collection for a nut that has an inner diameter slightly less that the required wheel tread.

Grip it in the three-jaw (perhaps with a spacer), so that the internal thread can be turned away and the hole opened up to match the wheel diameter. Face off both sides. Then, gripping the job inside the opening, take a tool that has been ground to a very narrow parting-off profile and cut into those hex lugs so that narrow slices of them project from a round shoulder. The actual amount you turn depends on the profile of the shoes you are modelling.

It is easier to drill the pivot holes before cutting the ring into individual segments. The end result is six identical, delicately machined brake shoes that didn't need any actual delicate machining – an excellent state of affairs.

Turning the inside diameter of a large nut to match a wheel.

MAKING BRAKE SHOES *continued*

The nut is then gripped on the inside.

Facing down to thickness . . .

MAKING BRAKE SHOES *continued*

. . . produces a finished blank.

A very narrow parting-off tool cuts into the blank to leave projecting 'ears'.

When the ring is sliced up it produces tiny brake shoes.

MAKING AXLES

You would think that turning a plain axle is a simple enough job. And really it is . . . except that the shoulder turned on each end, on which the wheel fits, has to be absolutely concentric with the main part of the axle itself. LBSC, for his larger models, recommended the wonderfully simple practice of chucking some round steel in the three-jaw. This could well be inaccurate as it stood, so he adjusted it with pieces of cigarette packing foil, between various jaws and the job, until it was spot on. This is a bit time-consuming for a beginner but it works well. It was my method of choice years ago. If you have a set of collets, these lovely devices will make gripping an existing true piece of rod simple.

Because – probably like you – I don't have to turn up dozens of axles a week, I am quite happy to use the four-jaw chuck. The jaws are independent and can be very finely adjusted. I bolt a dial indicator onto the tool post and rotate the chuck by hand, observing the discrepancies, and slowly adjust them out until the rod runs true. The first time you try this it seems slow and fiddly, but you will soon get the hang of it.

When you have turned one end of an axle, you don't have to go through the full procedure for the next cut. Just slightly loosen two adjacent jaws and slide the rod out. Put it back the other way round, then tighten the loosened jaws back up against the unchanged ones. You should find that the rod, naturally, runs pretty true, although I still like to check it anyway. Repeat the exercise with the other axles. We shall again employ what is becoming a familiar principle. Turn the shoulders down to slightly oversize, lock the depth of cut, then just slide the tool in and out sideways. This ensures, of course, that every shoulder is exactly the same size as its friends.

There is a method of turning perfectly concentric axles for the beginner. Chuck some oversize rod in the three-jaw and support it with a centre in the tailstock. The trick is to turn the slightly wobbly steel down to the diameter of the axle and turn the shoulders down as well, without disturbing the chuck's grip on the bar. Thus all eccentricity is removed.

Making Crank Axles

If the locomotive has an inside cylinder or two, you will need to make a crank axle. There is a good technique for fabricating these that works quite well for smaller-scale models. The webs (aka journals) are made from suitably sized flat steel bar. For a loco with just a single cylinder between the frames, you need two. If it is a two-cylinder engine, then, naturally, four are needed. But whether it is two or four, make two extra – for reasons we will soon come to.

You will need to mark out a centre line and then drill two holes in each web. One should be for the axle. The hole for the crankpin may be the same

DIAGRAMMATIC VIEW OF FIRST STAGE IN MAKING A CRANK AXLE

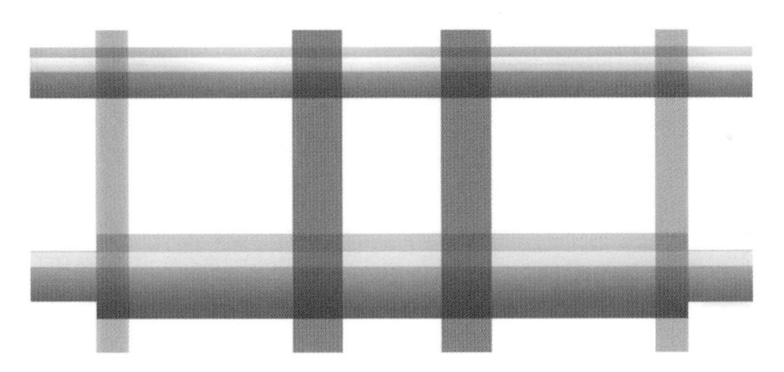

The assembled components of a crank axle. They will now be silver soldered.

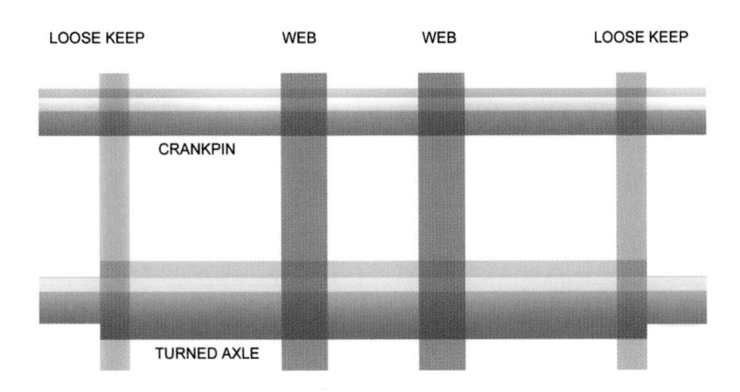

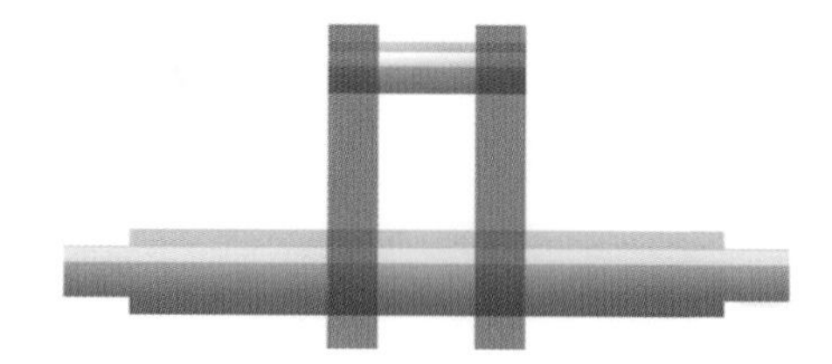

The surplus ends of the crankpin have been removed.

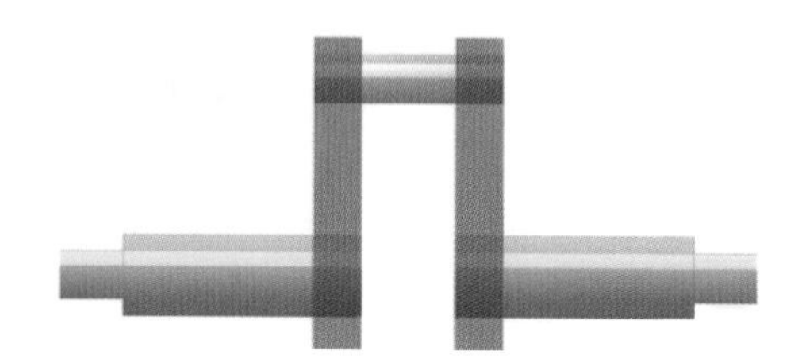

The central part of the axle is now cut away.

diameter or it may be slightly smaller, according to the job in question. The location of the holes must be identical in every web. I drill out the first web (holes very slightly undersize) and then clamp a second one tightly to it – and drill through, using the first as a guide. I drill out all the webs this way. I clamp all the webs together with a nut and a bolt through one of the two holes, and then check to see that the other holes are exactly true. If there is a very faint irregularity here and there, drilling out with the final-size drill makes that good. I then change over, bolt the other end and run the final-size drill through. The holes do not have to be a smooth, reamed fit.

Ignoring those two extra little webs we made, for a minute, make a slight countersink with a larger drill on both sides of all holes in the working webs. The rods and webs are assembled as in the diagram. Those extra webs are temporarily put on the ends, whereas two of the working webs are put in the correct position along the axle, as according to the drawing. This assembly is pushed down onto a true flat surface. It should be stiff enough to stay flat of its own accord. So it is taken to the hearth and the areas around the holes anointed with flux. Heat up the job to full cherry red and apply ordinary silver solder. Don't apply too much, as you will only have to get rid of the surplus afterwards.

ROUNDING WEB ENDS

The ends of the webs are simply cut at right angles. They work perfectly well and are pretty well out of sight. But if you feel the urge to put a radius on the ends, as per the full size, the two holes you drilled in them give you the means to do it easily. Take a tight-fitting bolt and a nut and run it through a couple of them. You will need to centre-pop accurately the hex head of the bolt. Put a couple of webs on and screw *very* tight with the nut. Catch the protruding part of the bolt in the three-jaw and then bring up the tailstock to push a centre into the bolt head. You can see from the stylized image how one web (thinner than it should be for clarity) would look. You can put two or four together on one long nut, then run a further nut and bolt in the other hole to hold them in alignment. The tool will only be cutting intermittently, so be very gentle with the cuts.

Finally, the webs are similarly bolted to the other holes and thus the other ends are cut. By doing this, the radius of the ends is true to the centre lines of the axle and the crankpin, and so look right.

One of those small blowlamps that screws into the top of a can may just about give enough heat for a small-scale loco, but life is easier if you have a separate nozzle that is connected to a dumpy bottle on the floor, via a hose. You would need one of them for most boilermaking anyway. If you are only planning on making one engine in the immediate future, perhaps it is worth finding someone who will lend you a decent-size blowlamp in exchange for a small consideration for gas used.

Having silver soldered the two axles to the two webs, it is time to saw off the surplus pieces of the crankpin rod, leaving the piece just remaining between the webs. Carefully file off any surplus silver solder. Then comes the magic moment. Saw off the main axle between the webs and carefully file the job down neat and flat.

A Double Crank Axle

If there are two cylinders between the frames, two cranks are required in the axle, at right angles to each other. Fortunately, this is easily achieved. Regard the single crank you have just put in as a plain axle for now. Lay it flat and then put on two

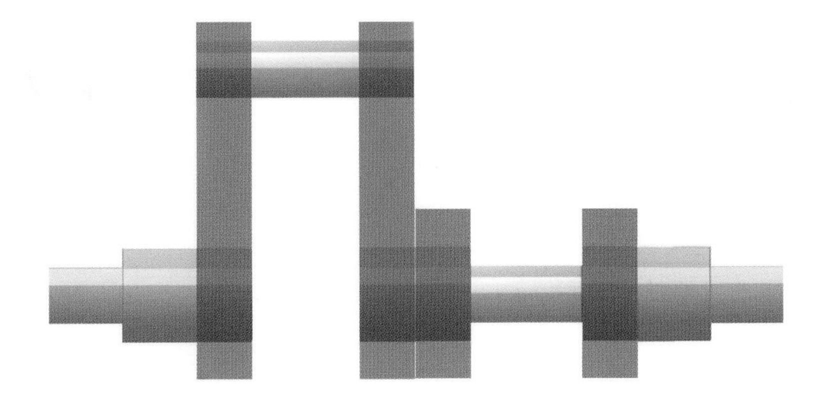

A double crank axle merely builds on what we have achieved this far.

more webs, sticking vertically upwards. Push a piece of crankpin steel through the holes. You will only be able to use one temporary web at the outer end. But if everything is nice and tight, the job will hold together truly whilst you silver solder these new parts together.

Strictly speaking, you should use a silver solder with a lower melting temperature than you used for the first crank, so as not to disturb existing joints. However, you can put a simple heat shield around the original work. Cut a square from a piece of white heat-resistant mat. Then cut a hole from one edge to the centre and wrap that cut around the axle. Get in quickly with as much heat as possible; do the new joints and get out again. The first solderings will remain undisturbed. Leave to cool naturally.

When this new work has been cleaned up, you now have an axle with a double crank in it. When all is cool, put an axle end into the four-jaw, set to run true. Pull the lathe over by hand and look for any hint of the job being out of alignment. It is just possible that the heating might have built up tension in the axle material. When you cut the gap away, the two halves could have microscopically sprung. Because you have made the axle from mild steel, you can place the job in the three-jaw and gently tap any irregularity out. Note that this is something you should never attempt with cast iron.

Eccentrics

A crank axle, or an outside crank, is used to transform backwards-and-forwards movement into rotary motion. But to do this in the reverse direction we can use something simpler. To use a turning axle to create push–pull activity, an eccentric will do the job. This consists of a disc of metal with some sort of groove cut in the rim. It is drilled off-centre and so wobbles as it rotates. There will probably be a small portion of it that is turned true to the axle. There will be a grub screw running through this, that will grip on the axle.

A good engineer may well turn the metal off-centre by holding it eccentrically in the four-jaw, but there is a simpler system that is popular. Place a cob end of mild steel in the three-jaw and face off the end. Turn it to the profile shown on your particular drawing, allowing something for the little concentric bush we will need in a minute. Part off the component.

Find the centre point of the disc and put a faint pop on it. Next, measure out towards the edge, for

An eccentric can seem a very small part to try to make at first, but our eyes quickly become accustomed to it . . .

(1) It starts with a plain bar.

(2) The bar is turned to profile.

(3) This is then parted off.

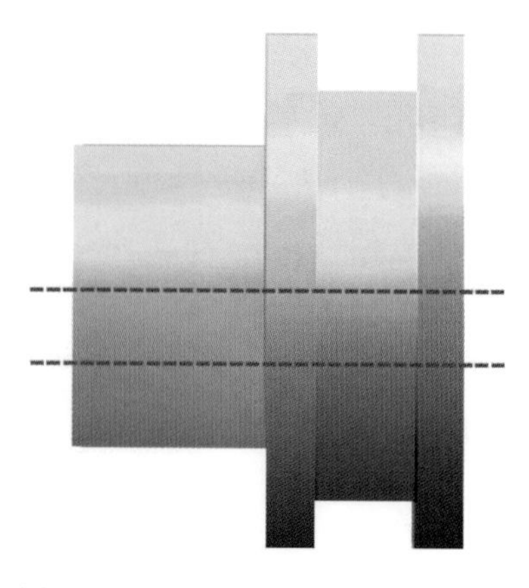

(4) A hole is marked with the right degree of offset . . .

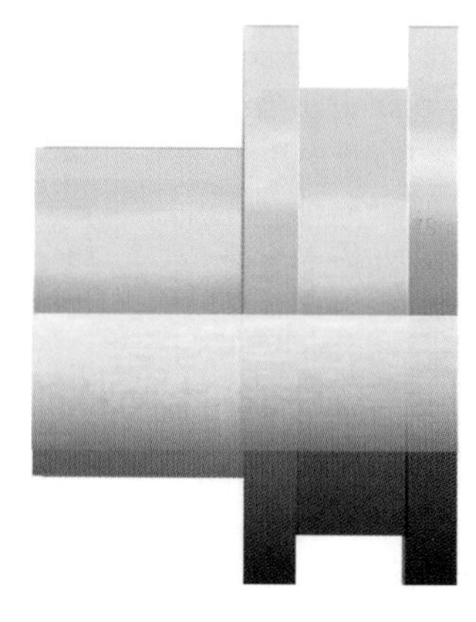

(5) . . . and then bored through.

(6) Finally, it is mounted on a mandrel and the profile of the shoulder is turned. A hole is drilled down into this and tapped. Thus a grub screw will now hold the eccentric tight to the axle.

a distance of half the amount of movement you want to transmit. Make a centre pop here and then open it with a Slocombe drill, using the bench drill. Follow this up by drilling the hole to size – this should be a snug but movable fit on the axle.

You will remember that we made a mandrel for turning wheels. Make another like it, but this time the turned portion will be to the axle diameter, rather than that of the little shoulder on the end.

Put the eccentric onto this and turn the little boss down to size. Drill a hole into this and tap a thread. Then insert a little grub screw. I like socket cap screws for this purpose. In practical terms, I start with a suitably sized grub screw I have in stock and then drill and tap the hole to suit!

You now have an eccentric that, in tumbling, imparts the movement you want. Most likely, this will be to operate the valves for the cylinders, but it is also how we get motion to drive an axle pump. By rotating an eccentric, which drives a valve rod, to different positions on an axle, the timing of admission and exhaust of steam are affected.

Eccentric Straps

An eccentric strap (sometimes called a sheave) needs to be a snug, but not binding, running fit over the eccentric – tight enough for there to be no play, but without the slightest hint of serious friction. To aid this, either the eccentric should be made of mild steel and the strap of gunmetal/phosphor bronze, or vice versa. The dissimilar metals, well lubricated, will slide smoothly against each other. They will also give many years of substantial service without undue wear.

The outside profile of the strap is mechanically unimportant. It can be shaped like the prototype, or it can be a plain square or rectangle. But what *is* vital is that the hole in the centre is perfectly round. And here lies a possible trap for the unwary. The temptation is to bore out a hole first and then saw the strap in half, but every saw cut has a thickness. This will make the hole smaller and also it will be very slightly out of true. However, there is a good procedure to combat this. Clean up the profile of the casting, if used, with a file. Face both sides down to the correct thickness in order to fit in the groove of the eccentric. If you are making the straps from mild steel sheet, it makes sense to have machined the groove to fit the plate.

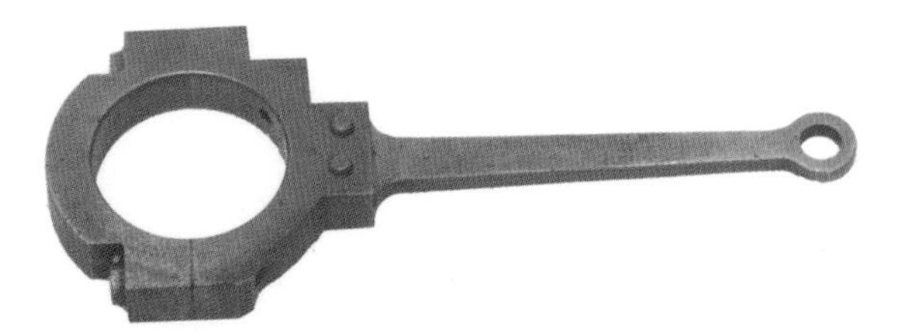

This assembly has been machined from cast iron. The lubricating hole is just visible inside the bore.

If you are designing your own simple eccentric straps, make sure that they are not too deep. They need to clear the tops of rails as the engine goes over points and crossings. There isn't a lot of room to play with if the locomotive has the small-diameter wheels often associated with narrow gauge engines.

Drill through for the screws that will hold the two halves together. The drilling can be done in a bench drill. The holes may be very small, using a very thin

A pair of steel eccentric rods is spliced onto gunmetal straps.

A gunmetal strap casting is an unpromising object.

Having been roughly bored and split, it begins to look more interesting.

A tiny eccentric strap – as it starts and as it ends up.

drill. These like to break for a pastime, if you are not careful, so use a nice high speed and keep pulling the drill back to clear any swarf from the flutes. If drilling in steel, a drop of cutting oil will help. Part of the job of oil is to cool; any oil is better than none, but a proper cutting oil is best.

Mark out a vertical cut in the strap and then make your very best efforts to cut a truly straight and vertical line. Make tiny single pop marks to ensure that you put the two halves of the strap back together the right way round. If you are machining two straps, make a pair of tiny pop marks on each half. This ensures that you don't mix them up. I know it sounds like I am labouring a point, but it is magnificently easy to get them muddled.

Screw the two halves together and then make a centre pop in the middle of the (tightly clamped) saw cut. Pilot-drill this and then we can think about opening up the hole to size. The final diameter of the hole needs to be to an accurate size. If you have a very full range of drill sizes, you might be lucky and find one that is right (make a test hole in a piece of scrap material first, before spoiling your strap-in-progress).

However, it is better to place the job in the four-jaw – adjusted so that the centre pop runs true – and open out the hole with a drill until there is just enough room to get a small boring tool in. Continue to bore out the hole until it is just too small. The next step needs care. It is very easy to take a fraction too much off and then have to start again.

Fabricating a strap from the solid starts by shaping a chunk of metal.

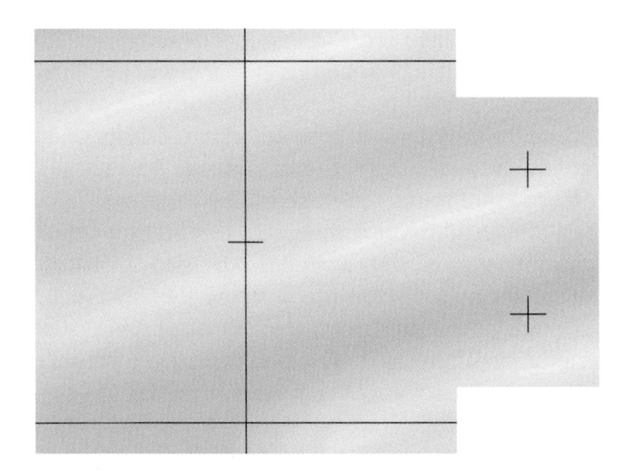

The next step is to drill through for the two connecting screws.

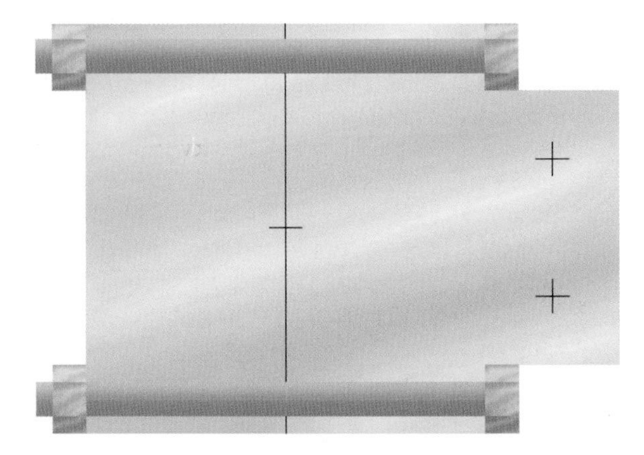

The block can now be sawn in two portions and these are bolted together.

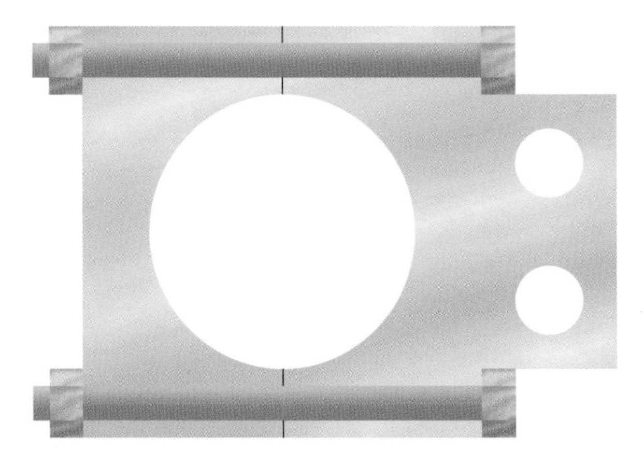

Finally, the hole is drilled and opened out to suit.

A digital caliper is a great help here. Just take the faintest whiff of metal from the job and then withdraw the tool. Take the strap from the lathe and offer it up the eccentric. A light wringing fit is fine when the nuts are tightened on the screws.

If you are using a casting, there is often an additional little 'flat' cast in. This goes on the top. You can drill a very fine hole through to the inside of the strap to allow oil to run down. Drill out a small countersunk pocket at the top of the hole to act as a reservoir.

Bedding In

If it is tight, clamp the eccentric in a nut and bolt and catch this in the lathe chuck, paint some brass polish onto the eccentric and then bolt the strap on. Cover the lathe bed with a bit of newspaper. Hold just the strap with a pair of pliers (card in the jaws to protect your shiny metal) and switch the lathe on to run fairly slowly. This will bed the eccentric and the strap together splendidly. If you have already fitted a connecting rod to the strap, this can merely brush against a piece of wood on the lathe bed, wobbling up and down. Don't go away and leave the job. This bedding in is quite quick.

The last job is to make the rod that connects the strap to the valve spindle. This will vary with every design. It may need shaping and bending to clear obstacles. A common arrangement is to have a U-shaped piece (a yoke) screwed to the valve spindle. Thus the distance between the valve and the eccentric strap can be adjusted. The flat connecting rod is bolted between the jaws of the yoke.

CYLINDERS

This is the big one – the job that calls for precision. Fortunately, it can be broken down into manageable tasks. We know already that there are two sorts of cylinder assemblies. Piston valves were favoured by many later prototype engine designs, but are said to be more challenging for a complete beginner to make in miniature form. We will therefore concentrate on slide valves in this particular book. Even so, we have a couple of choices of approach. The first uses castings as a basis for our machining. This can give a rounded shape, useful where fidelity to the

Machining a cylinder block involves a lot of facing off. This particular block happens to be for a piston valve unit.

real thing is called for. There is a huge range of sets of cylinder castings available. If you are working to a specific design, you can probably order them and may well also be able to get a full range of other castings, such as chimneys, grates and axle boxes.

If you are modelling something original, it is quite likely that you will be able to obtain a set of cylinder castings with a similar bore and stroke, in the scale to which you are working. For instance, although the interest in O gauge (7mm scale) live steam has diminished over the years, there are still many commercial designs and castings to be had. Such things have very long lives. I recently machined a couple of cylinder block castings that had been cast *sixty* years earlier! Given a choice, always plump for gunmetal (or bronze) castings rather than cast iron. The latter are always prone to corrosion and need oiling up between runs to keep them in good order. That said, it is no great hardship if you do have to settle for iron.

A completely machined slide valve cylinder, which started out as a casting. It seems complicated, but really is just a case of doing one stage at a time.

The second type of cylinder assembly is one that is purely functional. The job is machined out of rectangular blocks of metal. If it is an inside-cylindered model, the appearance is unimportant. With careful design, the combined width of the cylinder and valve chest can be such that they are exactly right for fitting between the frames as an extra frame stay. A single cylinder between the frames can have a bigger bore than is strictly correct, to make up for the fact that there is only one of them. Single-cylindered locomotives are not 'self-starting', but you can get a good punch of power from them. What's more, the beat of a single cylinder is slow. Even though it is technically wrong, it sounds lovely and crisp coming out of the model.

Machining Valve Chests

There are various ways of machining cylinder blocks and valve chests from castings. I will suggest a particular *modus operandi* that suits the smaller scales we are looking at. However, if someone in a local model engineering club suggests a different approach, then be open-minded about it.

Start with the valve chest as a good first exercise before tackling a cylinder block. The need is to produce an 'open rectangle', of which all outer faces are true and are at 90 degrees to their neighbours. Make sure that there are no odd lumps and bumps sticking up from the casting – a rare but occasional side-effect of the casting process. If so, file them off by hand. Thereafter it is a case of putting the casting in the four-jaw and machining all the flats. The truth of your lathe should ensure that everything is accurate. A smooth finish will come from using a round-nosed tool. At the ends of the valve chest casting there may be little extensions that will become bosses. Just machine around them, perhaps with a narrow parting-off tool to get true internal right angles. You might also like to drill a pilot hole for the valve rod, using a tailstock chuck, immediately after turning that end of the casting. This ensures that the hole is at 90 degrees to the face you have just turned.

A steam chest casting saves the labour of opening out the centre. Bosses are cast in.

PROTECTING FACES

Make it an unbreakable rule that whenever you have machined a new face, you protect it thereafter. Never grip it in a vice or chuck without simple packing; even strips of thin cardboard will do. Don't put a casting face down on a gritty worktop, and keep it somewhere that heavy metal bits cannot drop onto it. Indeed, it is good practice to put machined and cleaned castings into a box with a lid, with them sitting on a piece of oily cloth or tissue. It may be weeks or months before you get round to assembling them. It is good housekeeping.

In case you are wondering, the faces of the inside of the 'open rectangle' need not be machined. Provided that the slide valve can travel freely, the condition of these matters not one jot. And one other thing – don't drill the valve chest for the fixing holes at this stage.

Accuracy of Measurement

So far, we have described how to machine faces accurately. But, of course, we also have to think about turning something to an exact size. It can be done with measuring devices, like the tail that sticks out from a dial or digital caliper, but the lathe has its own built-in measuring ability. Each time a handle is turned to move the cutting tool, it will advance for an exact distance for every turn of the wheel. This will be either a metric or an imperial distance. Suppose we have measured the job and find that it is 1.3mm too long. If we know that turning the end screw advances the tool 1mm, then we know that we need to turn the handle one complete turn and then three-tenths of a turn. The handle may have a small dial marked with tenths anyway. In an imperial lathe, we would be looking at thousandths of an inch.

It also makes sense to consider turning both valve chests at the same time, rather than completing one and then starting the other. If we turn a face in one casting to a set distance, then back off the tool, say, six turns, by taking out the first casting, replacing it with the second, and then advancing the tool six turns, we know that both castings will have been machined to exactly the same size. It takes a certain amount of patience constantly to swap castings as you go, but it ensures that both end up identical. And, after all, there really is no rush, is there?

Machining Cylinder Blocks

We really are into the very heart of the model here. But there is no need to be daunted. By now you will be used to the fact that following a procedure will guarantee accuracy. Start by looking at the casting and file off any odd little bits of metal that are really standing up. In particular, make sure that the valve face is reasonably flat, even though it is still in as-cast condition. Incidentally, with all castings, just check to see that there are no odd grains of sand, in any crevices, left over from the casting process. If so, brush them away with an old toothbrush. Sand and precision engineering equipment do *not* mix.

Grip the cylinder casting so that the port face is turned outwards. Adjust it so that this face is truly vertical. One way of doing this is by resting a true block of steel on the cross-slide and then pressing it against the face. Then tighten up. You could also use a steel set square. Tighten up, then check one more time.

Use a round-nosed tool to machine the face true and flat. You will need to check that the tool is at the right height. If it is a bit off, it will leave a small pip. We want to get this dead flat. When the job is done, back off the lengthways screw a set number of turns – say ten. Make a note of that number. Change one cylinder over for the other one and make the required number of turns as you machine across this second port face. This means that the distance from the face from the opposite side of the cylinder has to be the same for both.

Boring

The first step is to mark out the centre line of the bore accurately. Don't rely on a cast borehole being accurately located. Tap a piece of wood dowel into the bore. Using the drawing for reference, measure where the centre line should be and make a pop mark. With a pair of dividers, scribe the location of the true bore. Whilst things are set up, scribe the circle that forms the centre line of the cylinder cover fixing holes. When you are satisfied that it looks right, put the dividers somewhere safe, without disturbing their setting. Tie a little label to them saying 'Cyl covers' – and don't touch them until they are needed for this job. Remove the wood dowel.

With our small cylinders, a moderate-size lathe will turn them in the four-jaw. Adjust them until the scribed bore diameter runs true. Then steadily bore the hole, very slightly undersize. The block will need to be packed out from the chuck slightly so that the boring tool will cut all the way through. Make a note of the setting of the cross-feed handle. Back off a set number of complete turns, replace the cylinder block with the second one and repeat to that same setting. This means that the bores have to be the same size. Finally, run a correct-size reamer through and withdraw it.

If there is only a small lathe at your disposal, the cylinder block will have to be clamped to the saddle. This will have T-slots in it. These can be used for whatever clamping arrangements you may make. Working out how to do it can be a three-cuppa problem, but there is always a way. You may have a machine vice that clamps on. Whatever solution you find, the job will have to be packed up for the correct height.

We now need a boring bar. This is a long piece of steel rod that runs from the four-jaw to a centre in the tailstock, or between centres with a driving dog to the faceplate. Somewhere in the middle, drill a hole for a small boring tool to go through. At right

As an example of machining in reverse, here an axle box stick is held in the tool holder and is passed across a rotating mill.

angles to it, drill another hole, tap it with a thread and put a suitable screw in. This screw should be countersunk into the hole so that it does not protrude above the surface of the bar when fully tight. Alternatively, make it a grub screw; cut the head off and file a slot across the top for a screwdriver.

The little cutting tool sticks out from the side of the bar until it cuts just the right diameter and is clamped firmly in place. Thus when the saddle carrying the cylinder block is moved along, the rotating tool bores it to size. If the original core hole was a bit off-centre, the cut will seem slightly eccentric to it. But as we gently continue boring, we end up with a slightly undersize bore in the right place. Once again, we finish off the job by reaming.

If the gods are really with us, the existing borehole is in the right place. The final bore can then simply be drilled out in the lathe or – making sure that it is set up truly vertically – in the bench drill. Run the drill through and back several times without stopping. That might even give you a good enough finish to avoid the reaming. And here I am going to be slightly controversial. Provided the bores are true and smooth – and both the same – it doesn't matter if they are slightly over- or undersized by a couple of thou. We will make the pistons to fit.

Cutting Ports

You will see from any drawing of a slide-valve cylinder that the port face has three slots cut into it. In the case of larger castings, these slots – known as ports – may well be cast in already. In some of our smaller sizes, ports can merely be holes drilled in place. So long as steam can be transferred between these ports by the valve, they will work. We are not worried about the nth degree of efficiency. Our small models will do the job they are asked to do, quite happily – and will not be too worried about how they do it. But if you are aiming for rectangular ports then the easiest way is to use a milling cutter that cuts the slot by moving the job in a travelling machine vice. Make sure that everything is bolted rigidly. If this is to be your first attempt at this job, it would be a good idea to practise on that long-suffering piece of scrap material first. This will let your fingers get used to the feel of the job, before inadvertently wrecking a part-machined cylinder.

Small ports need to be cut adjacent to the cylinder ends. A small hole drilled into the ends of the casting, adjacent to the bore, is a good starting point. You can then file a broad nick to allow it to connect to the bore itself. Be careful not to let the

INDEXING

This is the engineering word used to describe measuring out points around a circle. For example, if we want to drill, say, four holes near the edge of a cylinder cover, to be equally spaced they would be at 90 degrees to each other. This could be achieved by drawing an accurate straight line across and then another at a right angle, but with a lathe, we have the perfect tool to do this precisely. If the job is held in a four-jaw chuck, the jaws of that chuck are very accurately aligned. So a simple device, like a small door barrel bolt, could be clamped onto a fixed part of the lathe so that the bolt could slide out and act as a stop for each of those jaws to bear against in turn.

Now for the cunning bit. Clamp a pointed tool sideways in the tool holder and, with the lathe stationary (but with a jaw bearing against the home-made stop), feed the tool across the diameter line scribed in the job and scratch a line. Repeat with the other three jaws against the stop and you have four intersections perfectly at right angles. If you only wanted three holes at 120 degrees, you could use the three-jaw.

For more intricate divisions you will have a whole lot of gear wheels at the left-hand end of your lathe. Count the teeth (mark your first tooth with a tiny blob of paint – you are bound to lose track otherwise!) and calculate how many teeth will constitute a certain number of degrees. You might even acquire one or two spare gear-wheel oddments that will fit on the end, which you then specially mark out for particular numbers of degrees.

Incidentally, the practice of cutting a fixed piece of metal by forcing a tool sideways across it is called planing. This is normally the province of much larger engineering tasks, although it has applications at small scales. If you were to plane grooves at 10-degree intervals along a stub of brass rod, it would give a non-slip finish that is easy to grip for unscrewing by hand, such as the cap of a displacement lubricator.

tip of the file scratch the bore of the cylinder, if it has already been machined. If in doubt, put a strip of card into it for protection. You will need to drill the steam inlet and exhaust passages. Those inlet holes are usually at a slight angle. I have got on best by cutting a notch of the correct angle in a block of hardwood and clamping the cylinder to it. This holds the block in just the right position for an ordinary bench drill to do the business. If you enjoy the luxury of a tilting drilling head, then use it. If you decide to use the tilt of a tilting table in your bench drill to get the angle, again I suggest that you just try out your ports and angles on a piece of scrap first – even a chunk of hardwood will do.

When both cylinders have had their port faces machined, lap them in.

Lapping

We can achieve perfection on a face by lapping. Typically, this means taking a small offcut of thick glass and putting some valve-grinding paste on it. The face is rubbed onto this in a continuous figure-of-eight motion, without any rocking. Because the glass is *very* flat, this grinds a beautiful true surface. Grinding paste comes in coarse or fine grade. I usually like to finish off with metal polish. Take care to rinse the paste away completely.

Incidentally, away from more advanced model engineering, this lapping surface also helps to improve the functioning of oscillating cylinders, especially those of Mamod railway engines. An alternative to the paste and glass technique is to place a piece of very fine emery paper, face up, on a true flat surface, such as the table of a bench drill (I use the heavy chromed plate from the bottom of a failed domestic electric iron). Never contemplate doing any grinding anywhere near the bed of a lathe, and really wash the job thoroughly in warm soapy water afterwards.

Machining Cylinder Blocks from Solid

For inside-cylindered engines, where the external profile is not important, we can machine cylinder blocks and valve chests from solid material. A 2.5in gauge loco might call for a starting block of metal (most likely hard brass or gunmetal) 30mm square. If we are lucky that block may really be square when we get it. So, hopefully, there are four faces that have already been machined accurately for us; it is just a question of sawing and then turning to length in the four-jaw. In this instance, the existing good faces should be protected right from the start. If the block is held in the vice, it does so with packing, or with smooth jaws, where available. Incidentally, as your vice jaws will probably just be bolted to the vice, it is no big effort to make your own smooth jaws. Unscrew the existing ones. A couple of lengths of clean, mild steel, cut to length and drilled for fixing screws will soon do the job. If you can get hold of suitable strips of hard nylon-type plastic, make a set out of that too.

With the ends of the cylinder blocks turned square, the position of the main bore is marked on one end. As with cylinder castings, this bore needs to be absolutely true in all planes. So we take care to get this right by using the methods described above for work on a casting. Start by drilling a moderate-size first hole and then open it up with successively larger drills. If possible, increase the drilling size to just a fraction under the required diameter and then finish with a reamer. But if the need is for a really large diameter, then it will be better to open the hole out until a boring bar can be used. The rest of the machining follows the same methods as previously described.

Machining Valve Chests from Solid

Only a couple of points need be made here. Again, if you are lucky, you may get a piece of suitable bar that has some true faces to start with. The rectangular opening in the centre calls for drilling holes and filing. It's a dull and slow job to do, but it only needs doing once.

Making the Cylinder Covers

These are usually round discs of brass or maybe castings. In the case of the latter, they may have little spigots cast onto them to hold in the chuck. By now, though, you will be used to simple turning jobs. If you have four cylinder covers to turn, consider the possibility of turning a cob end of suitable brass to diameter – plenty long enough to allow for them to be parted off later into separate items.

You have those dividers, still set for the diameter of the fixing holes. Now is the time to retrieve them and scribe a circle at the end of the cob. Using the indexing and planing described earlier, mark out for the cover fixing holes. Take the cob out and drill those holes with the bench drill. Your drill size should be the tapping size for the thread you will use, not the clearing dimension.

Then part off four discs in the lathe. Now what I should tell you to do is to use a parting-off tool, but this generates a lot of swarf with the width of the cut. So whilst no one is looking, use a hacksaw on the rotating job. Give each individual piece a generous thickness. Protect anything that the end of the hacksaw might bang into. Don't force the blade down with hard pressure. I like to use just the front end few inches of the hacksaw blade. This leaves the middle of the blade sharp for other jobs. There is also a bit more control somehow. Let the blade do the work. You will get a really narrow cut fairly quickly.

When all four discs have been parted off – and you have guiltily hidden your saw away – you will need to turn the outer faces flat and the inner faces with a small ledge that just fits in the bore of the cylinder. This should not be too deep, otherwise the end of the piston may bang against the covers. You want just enough to create a register for the location. Dimensions will often be given in drawings.

Drilling for the Piston Rod Gland

There is one more item that has to be very precise indeed. The piston rod has to run exactly along the centre line of the cylinder, so the hole drilled in that particular cylinder cover has to be precisely located. I like to adjust the cover so that it turns perfectly truly in the four-jaw and drill thereafter. We said a little earlier that we needed to turn a small ledge to locate the cover onto the bore. If you stop to think about it, the hole for the piston rod needs to be in alignment with the centre of that ledge. It therefore makes sense to drill it whilst the cover is still in the chuck after the turning process. You might even consider turning a tiny dimple in the centre of the cover, then when the Slocombe drill is used first, it will locate at that exact centre point. Thereafter an ordinary drill can actually drill the pilot hole. We won't drill for the size of the piston rod itself. The pilot hole will be opened out later when we make a piston rod gland.

To summarize: there are many jobs in building a small steam engine where we can get round problems with all sorts of little dodges. However, the bore of the cylinder has to be true and the piston rod has to go exactly down the centre line of it.

This is the one area where we take our time to get it right.

MINOR TURNING JOBS

By now, you will be familiar with how to read drawings and to interpret them in metal. For example, you will see that a gland is a kind of screw-in plug and socket, with a hole through the centre. There will be a recess for an O-ring. Making one involves no new techniques. Like so many jobs, the only thought required is the order in which to do the various turning and threading processes, and how to hold the parts. There is a variety of these small parts to make – things like buffers, either unsprung or sprung. It would be repetitious if I were to go through every one that you might encounter, but, to give an example, we can look at making a displacement lubricator together (*see* page 66).

This consists of a tall, thin container with a screw-in top that is steam-tight. Through this runs the main steam pipe to the cylinders. It makes sense to start with a piece of thick-walled brass tube of the right diameter (or fairly close, it isn't critical). A disc of thin brass is hard soldered to one end. For a convenient draining arrangement, buy a ready-made cylinder draincock from a 5in gauge engine. Drill and tap a hole in the base of the lubricator to suit. You then merely have to flip it open to drain out the condensate. A lesser arrangement is to drill and tap a hole in the side of the tube near the bottom and fit a small screw with a soft washer.

Sometimes, in an old set of castings, you will encounter things that are more trouble than they are worth. It would be just as easy to make coupling and connecting rods from mild steel as use these castings.

A hole is drilled through the barrel, both sides, not far from the top. Make this far enough down so that it isn't blocked by the screw-in cap. A thread is run down the inside at the top for, typically, 6mm. This is where a plug tap is useful. Ideally, you want to be able to run the full size of the thread right down to a particular location and then stop, rather than having it taper off. In truth, this won't make any difference to how well the lubricator works – it is just neater engineering.

The cap is a simple turning job, with a matching thread, to screw into the barrel. If you can knurl the larger-diameter top part, it makes it easier to grip. Alternatively, you could plane grooves as described elsewhere. This top is mostly going to be unscrewed with a pair of pliers anyway. It makes sense to saw a line across the top so that the cap can be undone with a screwdriver. A neat little dodge is to silver solder a piece of thin steel rod across this groove and then bend one end up at right angles. Cut to length and you have turned this purely functional lubricator into at least a vague appearance of it being the handbrake standard and handle of the prototype.

Put the completed assembly in store until it is time to erect the engine. When that happens, the steam line is threaded through those two cross holes. When the lubricator is in its intended location, peer down the top of the barrel and you can see the steam pipe going across it. With a long pin, scratch the pipe, midway between the sides. Take the barrel off and then drill a very tiny hole into the pipe, where the scratch mark shows. The amount of steam that will condense into water through a hole depends on how hot the steam is. This can vary by how far the lubricator is from the boiler and hence the length of cooling pipe if it is not lagged. A proficient model engineer will be

Here are the machined components to make a safety valve. By now, you will be able to work out how they were made from raw materials and the order in which the jobs were done.

horrified at the term 'a very tiny hole', but if the hole is too large, there will be no real harm done – the engine will just use more oil. Equally, it will be difficult to make the hole too small, but a size to aim for could be size 50 in the old number series.

At this stage, we can start to think about building the mechanical parts of our engines entirely from scratch and/or using a few castings. But if we want to put together our own boilers too, then we should take a look at boilermaking first.

CHAPTER 8

Boilermaking

A locomotive boiler is a pressure vessel. Such things can go bang, although in practice it is very rare for them to do so. Usually when a boiler fails it is because a small hole has appeared at a joint, in which case it simply leaks. However, even this is not good enough for us. There are some steps that we can take that will remove many of the possible causes of failure. The main one of these is only to make boilers in copper (or possibly brass as a second choice). We will *not* make boilers out of steel, iron or tinplate.

The tube used to make the boiler will be solid drawn. It will not have a seam in it. There have been many commercial (albeit often toy-like) boilers put together with soft solder – particularly those with a very low boiler pressure. But I invite you to be resolved only to use silver solder. This makes a stronger joint and it will not melt if a boiler runs dry and becomes overheated.

Any boiler that gets built will have a safety test performed to ensure that it is suitable for going into a model steam engine. Some groups insist on a boiler certificate before allowing them to be used on group tracks and public occasions. Fortunately, these associations often have their own network of boiler testers (*see* Appendices).

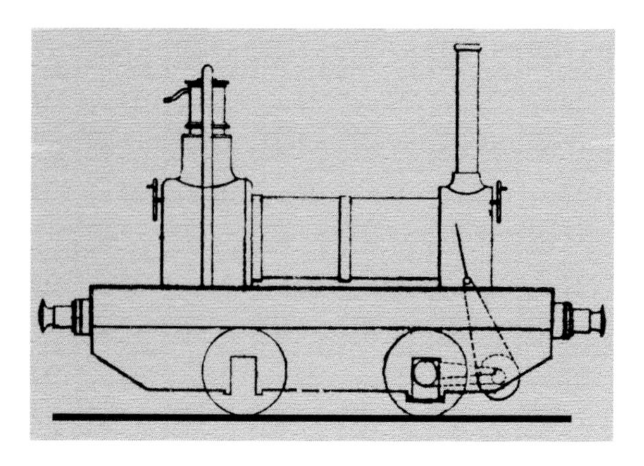

But there is no need for apprehension. A simple little pot boiler is an easy thing to build and should not cause problems. What we are talking of is a length of copper pipe with a copper cap at each end. Along the top will be a couple of holes. One will be a filler that has a screw cap. This cap can also be the safety valve. There is also a 'take-off point' for the steam to be lead elsewhere. With a fire underneath, the water boils and, being contained, builds up pressure. That pressure is used to push the piston in the cylinder. Simple. It can be seen on the most basic Mamod stationary engine.

MAKING A POT BOILER

Starting the Boiler Barrel

Take a length of tube, of the appropriate diameter, and cut to length with square ends. Actually, that is not entirely straightforward. If you have a large rotating pipe cutter; fine. If not, think about marking out the first straight line by wrapping masking tape round, adjusting until it looks right. Another ploy is to support the tube vertically on the bench by some devious means. Then, taking a suitable block of wood with a nail tacked on top, scribe round the tube. Finally, hold the tube very lightly in the vice and gently saw around, rotating the tube, rather than trying to cut right through in one go. Use a hacksaw with a fine-toothed blade.

With the boiler tube accurately cut to size, stick a piece of masking tape along its length (*see* page 34).

A very basic commercial boiler, but with the addition of a fire tube.

Draw a true straight line along it (a biro is a perfectly alright) and mark locations for any bushes. Drill pilot holes for these – lightly please, to avoid denting the tube by using excessive force. Then gently open the holes out to the required size. My own preference is to drill just undersize and then open the last bit out with a tapered reamer until the bush just slips in. This way, even if the bush isn't quite the right diameter I know that it is a good fit. This reaming also gets rid of the slightly rough edge that a drill might produce.

Boiler Ends

We next need a couple of copper discs for the ends. In the first instance, these will be plain discs. Now, in theory, it sounds simple: 'Cut two discs out of sheet copper.' But if you haven't done it before, it may not seem that simple. Yes, the disc shapes are marked out by dividers, but the perfect circle can't be cut out with a handsaw or a pair of tinsnips. Instead, I cut outside the circle and then patiently file down to it, by holding the disc flat on the edge of the bench and lightly filing sideways, rotating as I go. Occasionally, I will offer up the disc to the boiler barrel. Eventually it will be a moderately tight press-fit in. I then do the other end likewise. This may sound long-winded. You could sandwich a pair of slightly oversize discs between a pair of turned steel discs and file the copper down to them. But gently filing by hand until a good fit is achieved is probably as quick and simple. What's more, it reliably gives a good fit. Try to make the filing as accurate as possible.

An alternative is to use a hole saw. If you are exceptionally lucky, one will be the perfect size. If not, use the next larger size that you have and, when the discs are cut, file down to a template from there. If you go about using a hole saw the wrong way, all sorts of disasters lay in wait, but there is a safe method. Start by screwing two oversize pieces of copper to a backing board. Fix them at the corners in what will be the waste areas. Securely clamp or bolt the board to the table of the bench drill. The hole saw will have to be used without a pilot drill. Depending on the pattern of your nest of saws, this may mean making a short shaft with a dimple in it. This wasn't needed by the example in the photograph. There are tables of speeds for using hole saws on copper, but it is easier just to put the bench drill on the slowest possible speed and gently feed the hole saw into the job. Be patient and don't rush it.

Put the boiler barrel and end discs into a pickle solution overnight. Thoroughly rinse thereafter.

Making the Boiler

Press one of the discs into the barrel so that it is recessed by about 3mm. Set the barrel down on the hearth, with the disc facing you. Pack the boiler around with refractory material so that the tube is in a slight recess. This will be a simple silver soldering job, so can be achieved with a plain stick from a hardware store. A generous tablespoon of borax is mixed with a couple of drops of water to make a stiff paste. With a small paintbrush, paint this paste round the recessed rim of the disc. Try to form a fillet just around the actual joint, without smearing over the rest of the face of the disc. It makes for a neater job.

Get the blowtorch roaring (the little blowlamp screwed to a small gas canister may not be strong

A boiler tube with the ends cut square. The smaller fire tube has had cross water tubes silver-soldered in place.

A hole saw, without the pilot drill, can be used to cut discs in copper sheet, provided it is securely bolted down.

enough) and make the entire end of the job very hot. If it is hot enough, the silver solder will instantly 'flash' and run round as a liquid when you touch the slight recess. Don't be fooled by a blob just falling off the end from the heat of the blowlamp directly on it. But you really can't mistake that 'instant liquid' effect of the silver solder doing its job. If it won't flash, the most likely cause is that you are trying to do the job with too small a blowlamp.

Lack of experience may make you worry that you haven't used enough solder and you will feed a bit more in. This won't hurt but will look unsightly. Leave the job to cool partly and then, using a pair of fire tongs, pick it out of the hearth and quench it in cold water. That lovely clean copper boiler end will now look terrible – a hotchpotch of different colours and with excess solder disfiguring things. That is normal and it will clean up later.

Out of curiosity, hold the boiler with the soldered end at the bottom and pour a little water into the open top. You will be delighted to see that it does not leak at the bottom. Time for mild self-congratulation, before repeating the job at the other end.

A refinement of fitting boiler ends is to have them softened and then hammered around a former to form a flange, but properly hard soldered edge joints are more than satisfactory for our needs. This is especially true because of the boiler end plate being pushed slightly in from the end of the tube. When hard soldered on the outside of the end plate, a tough ridge is formed that reinforces the 'stiction' of the soldered joint itself.

Fixing the Bushes

Finally, lay the boiler onto the hearth with the bush holes along the top. Anoint the bushes with the borax flux and push them into the holes. Pack refractory material around these holes, leaving the newly soldered boiler ends uncovered. Heat the holes up and apply the silver solder. You really won't need much. There is no need to worry about melting the joints with the end discs. It would take quite a time to get them hot enough again to be vulnerable – you will have soldered the bushes long

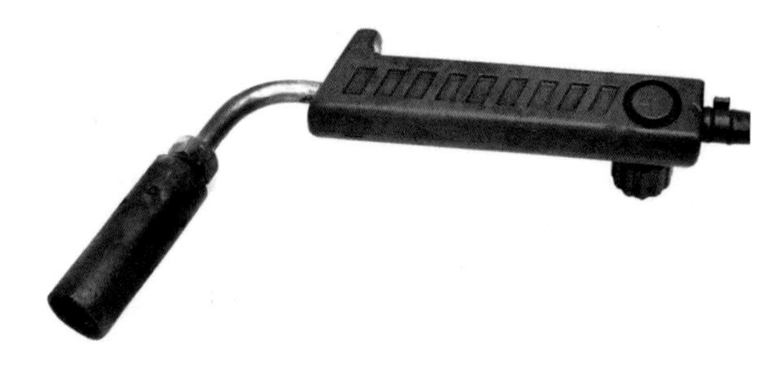

Boilermaking may call for more heat than a little canister blowlamp can provide. Here is my trusty torch, straight off the hearth.

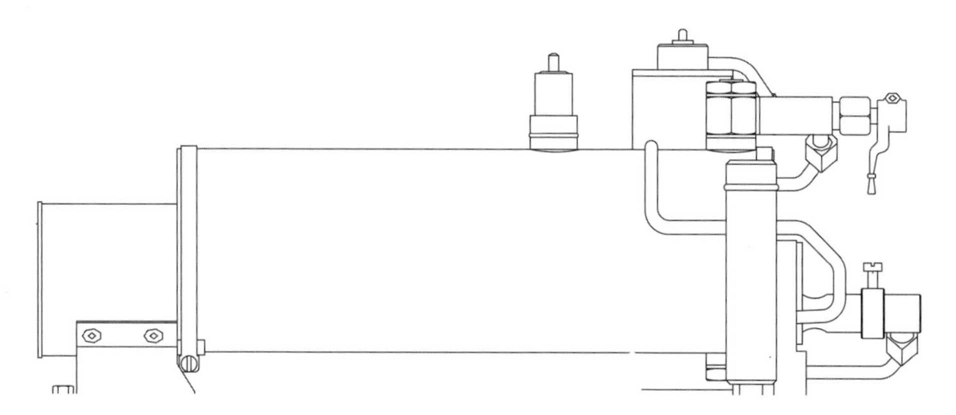

The outline of a gas-fired boiler, complete with ancillaries. (Courtesy of Roundhouse Engineering)

before there is a problem. If you are worried, rest a chunk of metal on the top corner of the boiler, so as to conduct unwanted heat away. But it really won't be needed.

When it is cooled down, pop the assembled boiler into the pickle solution again, rinsing off in clean water the next day.

Provided you follow the above steps, you should have produced a perfectly sound and strong boiler. There may be some excess solder here and there, but that is purely a cosmetic thing. If you have a reliable safety valve already to hand, you could put some water into the boiler and then screw it into the bush with a red fibre washer. Blank off any other bushes with appropriately threaded plugs, again with washers, and then prop the boiler somewhere. The part-opened jaws of the vice will do. Apply a gentle heat – nothing too ferocious – from a blowlamp. After a while, the safety valve should lift and you should see no little weeps of water elsewhere.

This practice of having a simple steam test before a hydraulic test should only be applied to small, low-pressure pot boilers. A better practice is to make an adaptor fitting for one of the bushes so that you can pump up the air pressure with a car foot pump – whilst the boiler is sitting in a bucket of water.

A refinement of the plain pot boiler is to have a couple of tubes looping down along the underside. Not only do they form an additional heating surface, but they also promote water circulation. There is often a friendly bubbling sound as the water starts to heat up. However, there is a problem. A naked flame needs height in which to develop its full potential. This presents no difficulty for a boiler in a stationary engine, but model railway engines have to follow predetermined dimensions and there may well be insufficient space for us to enjoy these tubes in small-scale railway engines.

FIRE TUBE BOILERS

A simple pot boiler described above will either have a spirit burner or a gentle gas poker underneath. The next step would be to run a plain fire tube from end to end. This lets a jet of flame, from a small commercial gas burner, run through the boiler to heat the water. To build a boiler like this simply needs the end plates drilling to accept a length of copper tube of appropriate dimensions. This is hard soldered in place at the same time that the end plates themselves are soldered up. This sort of fire tube is perfectly adequate for most needs and is used commercially. If we wanted to extract a little more heat from the flame, the fire tube could be slightly crimped inwards a couple of times to make the gases gently swirl. A very simple dodge is to slide a long, lightly coiled length of copper wire into the tube. This glows red hot and so becomes a radiant heater in its own right. But it also makes the gases swirl and thus delays them slightly as they pass through the tube. Finally, we could hard solder several cross tubes – typically

3/16in diameter – into the fire tube before it is soldered into the end plates.

A variation would be to solder several copper rods across the fire tube instead of the little water tubes. These rods would extend outside of the fire tube a short way. The rods get hot and this heat is conducted into the body of water in the boiler by the extended ends. It all helps to extract more heat from the fire before it passes up the chimney. Take great care not to overdo any of these refinements; there is an overriding need for that jet of flame to be able to burn uninterrupted. If it is too choked off, combustion will be incomplete. The roar will sound strangled and the flame may keep cutting out. As we move up the ladder of boiler complexity, we will pass on from DIY store silver solder and borax, and enter the realms of Easy-flo and the correct flux from a model engineering supplier.

Having mentioned these refinements, it is worth repeating that a properly set-up fire tube and burner will generate all the steam you are likely to need. However, we shall meet such refinements again with more advanced spirit-fired boilers where the hot gases move through much slower anyway.

SMITHIES BOILER

This consists of a long boiler that can be quite small in diameter. It runs inside a dummy outer shell. At the open rear end is a source of heat. This is usually in the form of spirit wick tubes, but a small open-topped gas burner could be used. The heat is conducted inside the dummy outer shell, enveloping the inner boiler and emerging out of the chimney. Because the height of the chimney above the flames is only 25mm or so, instead of the much greater height of the real thing, there is no natural draught to suck the gases through. So, a little auxiliary fan is needed to draw the flames through, in order to get the engine into steam. My own refinement of this was to put a tiny fan and battery in the back of a tender and run a pipe to the engine, via a stub of flexible hose, and then on to the smokebox where it formed a second blower pipe to work in the normal way. With internally fired boilers like this, it is easy for a novice driver to loose steam to such an extent that the engine can't bring itself back to life without an auxiliary fan being stuck on top of the chimney. How convenient then, to be able to flick a small switch and shoot a fine jet of air out of the chimney and get those flames dancing again.

From this simple but elegant boiler design there have grown more refined developments, which have advanced the principle of putting heat into the open bottom for a firebox area and then allowing it to make its way to the smokebox. Particular types of these are well known in Gauge 1 circles, and Aster has used them to great effect in some of their products.

PERIPHERALS

Superheating

We have said that the gases that steadily pass through fire tubes – as opposed to gas jets – can be slowed down and exploited further in various ways. In addition to those mentioned, we can consider superheating. When steam is taken out of the boiler and through the regulator, it makes sense to run the steam pipe through the flames slightly to then superheat it and thus dry it out. This gives it a bit more fizz in the cylinders. With a pot boiler this simply means running the steam pipe above the wick tubes. The effect can be heightened if the soft copper pipe goes through the flames and then bends back on itself . . . and then again . . . before going into the cylinders. This means that the steam has been through the flames three times and is getting nicely hot. I devised a neater and more efficient external Compton superheater, which is easily made. It consists of a length of 3/8in diameter brass tube. At each end we make a shouldered plug that has a hole 1/8in diameter (assuming a steam pipe of that size). The pipe bringing the steam in from the boiler goes through the hole at one end and is pushed through the tube until it nearly reaches the other. The exit pipe is pushed through the hole at the downstream end, again so

Case Notes:

Coal-Fired Boilers

We have agreed that building a coal-fired boiler is not for beginners – not impossible, but provides a lot of scope for frustration and disappointment. It is, though, possible to buy a ready-made coal-fired boiler. DJB Engineering currently offer such a boiler that will nearly be a straight drop-in replacement for several simple commercial models. I have also been impressed by the replacement boilers of John Shawe, built to order. Because of the work involved in building them, they are not cheap, but they would be a way for a determined beginner to enjoy coal firing.

To give a feel for what is involved in a coal-fired boiler, here is one used in Dante, a simple 5in gauge design of mine. There is a separate firebox fitted inside the boiler. Because this has flat surfaces under pressure, it has to be strongly reinforced with stays. A number of fire tubes run to the smokebox end. As the scale of the model gets smaller, the number of tubes decreases but they become larger in diameter, as a proportion of the boiler diameter. Occasionally, in Gauge 3 and below, you may encounter a slightly simplified form of firebox where the back of it is actually the back of the boiler as well. This is called a dryback boiler.

Because you can't scale down nature, the small size of the firebox means that a fire is not stable for long periods of time and needs careful attention. If it is at all possible, a wise builder will try to find ways of making a firebox overscale without it being visually obvious.

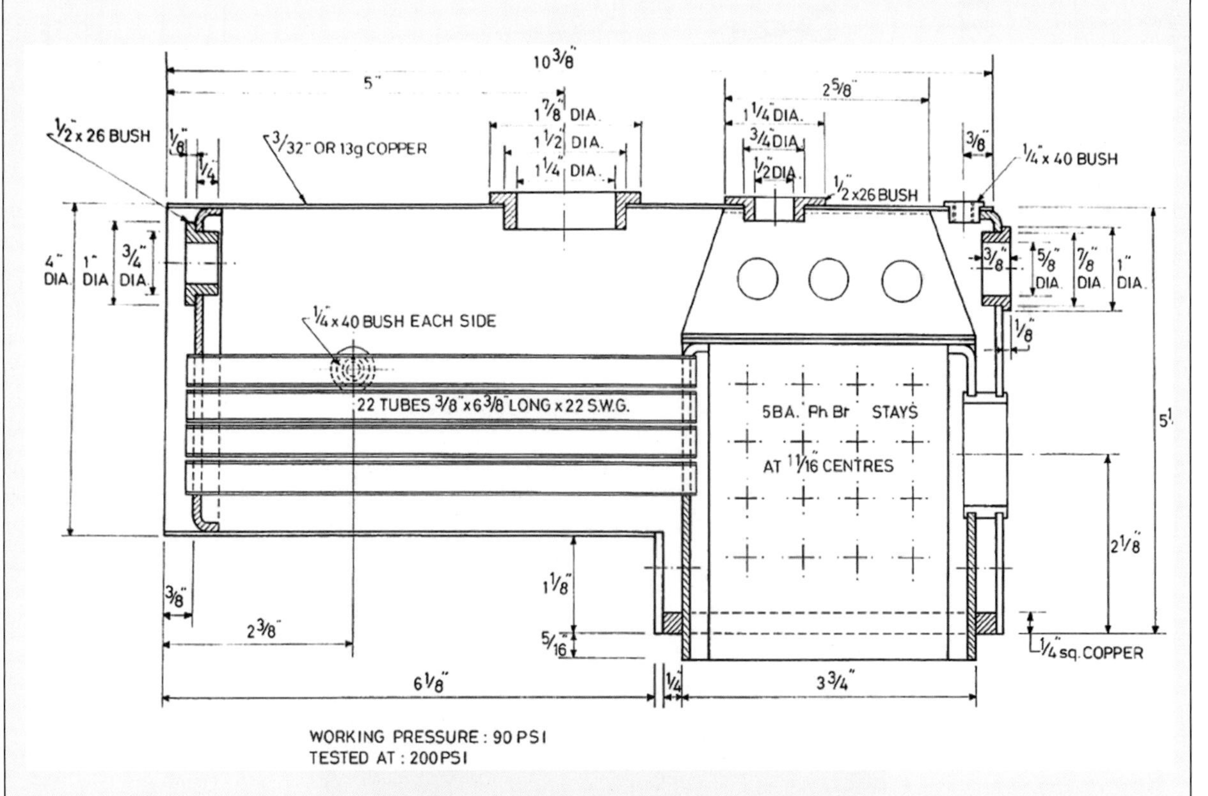

A coal-fired boiler in 5in gauge. It couldn't simply be halved in size for 2.5in gauge. There would need to be fewer fire tubes but of proportionally larger diameter.

Coal-Fired Boilers *continued*

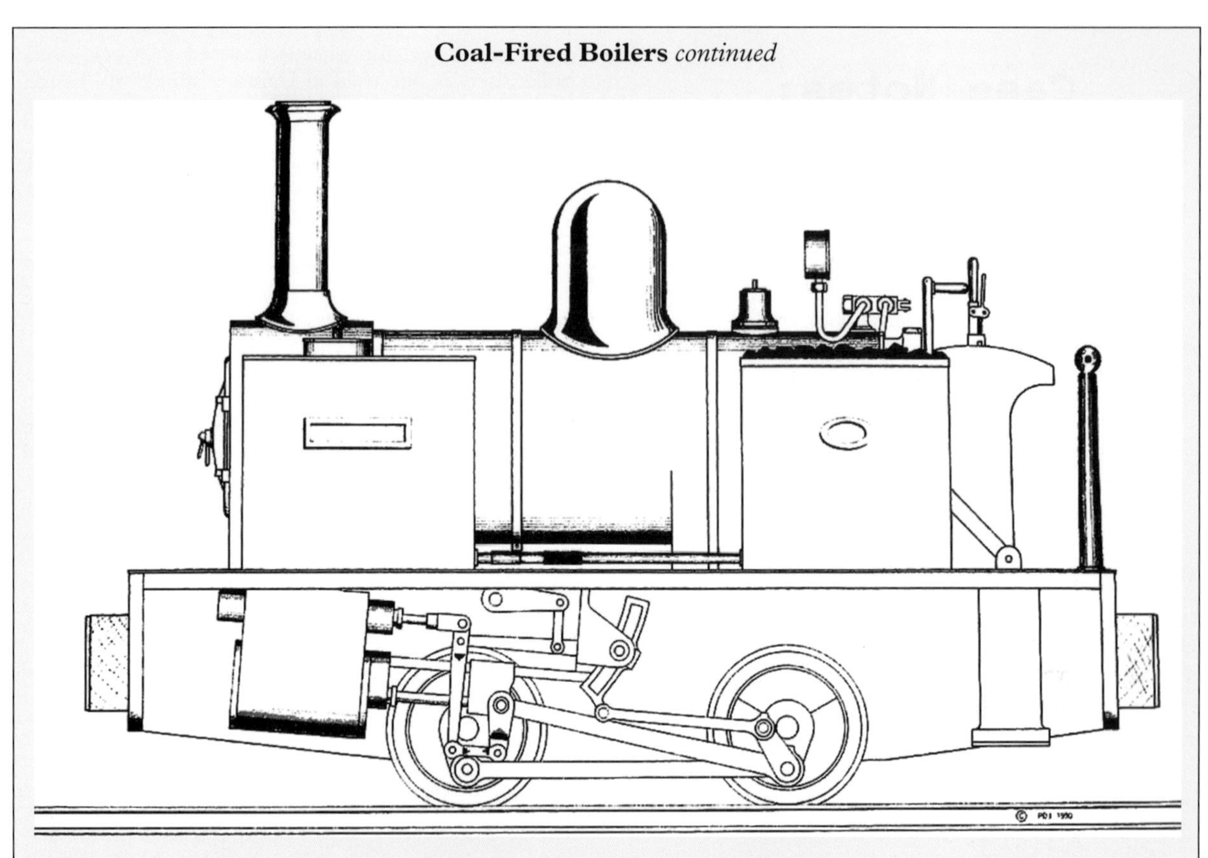

Dante was a 5in design of mine that was meant to be a compact 5in gauge engine. Its ancestry dates back to the primitive Gnat, as drawn in Chapter 5.

This design would make a nice basis for a model of the dainty little Dougal, to be found at the Welshpool and Llanfair Railway. Note the difference in size with the locomotive it is coupled to.

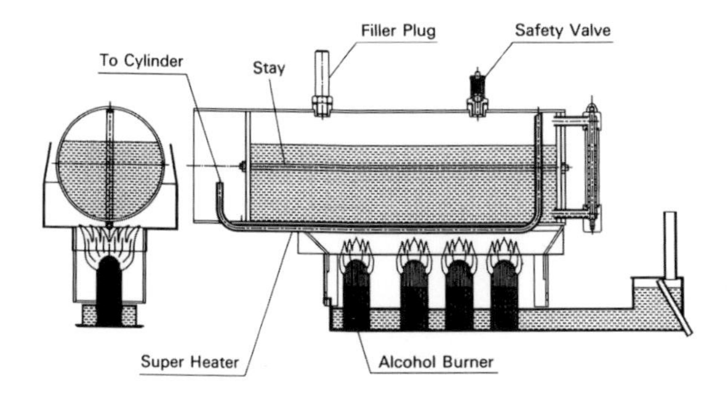

A simple pot boiler in diagrammatic form. Note the side shields. (Diagram courtesy of Aster Hobbies UK).

that it runs nearly the full length of the brass tube. This means that, once more, the steam has had to travel through the heat three times. But that brass tube also builds up heat itself, thereby adding to the superheating of the steam.

There may be a problem with finding free space in which to fit the burner in-between a crank axle and narrow frames. Quite often, a steam pipe will have to twist and turn in a convoluted route to clear everything, with the result that there may not be space for a Compton superheater.

Advanced Superheating

With a boiler that features a firebox and some fire tubes, prototype practice can be followed. Steam is taken along to the front of the engine, externally for convenience. The pipe then goes back into a flue tube, through the boiler, where it U-turns and comes back to the smokebox, to go into the cylinder/s. That very sharp return bend is achieved by running the steam pipe into a small piece of copper shaped like a spear head. A second pipe emerges, taking the steam back from whence it came. A refinement of this is to make the pipes so long that they reach right back into the firebox and the slug of copper glows hot in the flames.

Overall, we are not too worried about economy and extra efficiency in our small models, although some form of superheating will mean that the exhaust is drier and so gurgles and spits less. It may also give the engine's performance a little extra pep – but it is not essential.

The Boiler Level Plug

In the absence of a water gauge, it has been the practice of some commercial loco-builders to put a little bush in the back end of the boiler, at the height of the correct water level, so that the boiler isn't overfilled. Typically, three-quarters full is about right. I have been happy merely to peek down into the top filler opening until I can see that the level is about right, but a little bush and screw-in plug is cheap and simple to make.

However, we can take things further. Instead of that little screw, a small commercial valve can be screwed in. A pipe from this can run down through the cab floor to just above rail level. Any dribbling water or steam will therefore look like it comes from a prototypical injector. When the water has turned to steam, you will know that the water level was right.

One final refinement is to run the pipe under the running board and then down between the frames to just below the buffer beam. A jet of steam from here will look just like the effect of working cylinder drain cocks. On a cold morning, the cloud of steam coming from the front of the engine is most satisfying. The prototype Dacre featured this.

The principle of the Compton superheater for pot boilers.

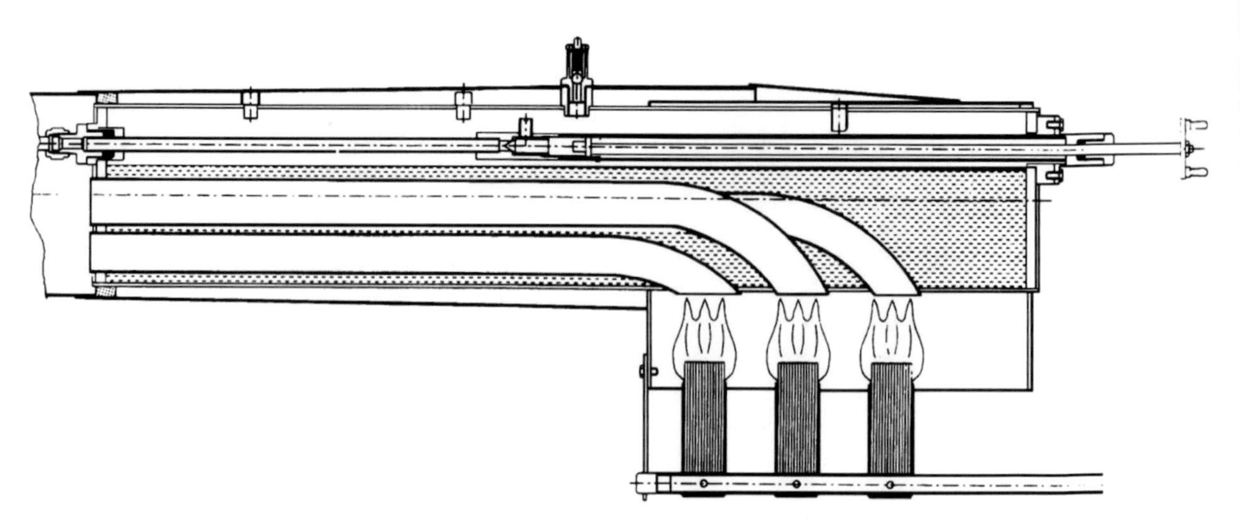

A spirit-fired boiler that has individual fire tubes: a further refinement. You will also note how the regulator rod runs the entire length of the boiler to operate it in the smokebox. (Diagram courtesy of Aster Hobbies UK)

Many of these things are interesting byways to explore. But that is what they are: byways. A simple pot boiler will produce the right amount of steam for the cylinders at the right pressure, provided it gets adequately heated. If there is a published design to work to, all well and good. Where a new design is undertaken, copying dimensions of something similar will give satisfactory performance for any given cylinder size and boiler pressure. If, despite all this, a new boiler turns out to be a shy steamer, there is usually some tweak to be had that will rectify it. Increasing the diameter of the wick tubes, or even squeezing an extra one in, often does the trick. With a boiler fired by a gas jet, putting a slightly bigger burner in place will rectify things. If there is going to be an error, it is better to produce too much steam than too little. But the likelihood is that, by following conventional design and practice, all will be well.

CHAPTER 9

Wirral: A Project in Outline

Having considered metalworking techniques, including machining, I would now like you to look over my shoulder as I think about, and build, a complete simple engine from scratch. I have chosen a 2-4-0 standard gauge tank engine in 2.5in gauge. But, in a sense, my actual choice doesn't really matter. The same processes would apply to any small steam engine. Construction is along traditional lines. Indeed, the origins of the chassis design can be traced back to an old LBSC design (*Rose* was a 2.5in gauge GER 2-4-0 tender loco, described in *Model Engineer* magazine in 1957). There are plenty of drawings and castings to be had, but, in this case, I'm putting together a design of my own, based on common practice. You will find that, when you work in a particular scale, you soon become familiar with typical sizes of things like axle diameter, valve rods and so on. For a guide to general practice, examine the drawing of Wirral. Even though this is a 2.5in gauge engine, it will give a clear guide to general practice with details. Read it in conjunction with the detailed engineering notes of Chapter 10.

There is a single cylinder between the frames that is also a frame stay in its own right. This drives a cranked axle and the timing is produced by a simple slip eccentric. A ready-made commercial cylinder assembly (such as Roundhouse) could be used, but I have put together a home-made one, hacked out of solid metal. There are numerous 0-4-0 tank engine designs to be had, so, just to ring the changes, I chose this wheel arrangement with the added interest of a double frame. Perhaps the most well-known example of this is the GWR Metro

tank. But don't read any significance into this; the design is merely suggested.

The first step in creating a design is to list some of the known dimensions. Plans of similar projects could be looked at to get a guide to these. But some things can be worked out by applying a bit of thought. The gauge determines the distance between the mainframes. We know the height and width (and distance above rail level) of the buffer beam from other 2.5in rolling stock. We also know the loading gauge – the maximum possible height and width of the 'envelope'. If we are following a specific prototype, the diameters of the wheels would be known, as well as the spacing between the axles (the wheelbase). Needless to say, these observations are equally applicable to an identical engine in Gauge 1 or O gauge. It is possible to produce reasonable working drawings from a couple of photographs, given a few known dimensions and some thinking time.

I will suggest that the availability of suitable wheel castings is of utmost importance in deciding that a particular locomotive is a 'possible' for modelling. Making spoked wheels from solid is not for beginners; even most experienced modellers tend to shy away from it. It is worth noting that where there is a blueprint for an established design, there are usually castings to be had somewhere.

BONDS COMPONENTS

Many years ago, Messrs Bonds produced a range of designs based around a limited number of standard components. The accuracy of the model sometimes suffered to accommodate these parts, but these locos were satisfactory performers. If you are really stuck for a particular wheel or cylinder casting, make enquiries about Bonds standard parts. There are still large numbers of them sitting in drawers around the country. Your relevant scale group would be a good starting point for your search.

The decision has to be made as to whether to make this a sprung locomotive or not. If it has suspension, rectangular cut-outs will be needed in the frames for the axle boxes to slide up and down. If there is no suspension, we need only drill holes in the frames for the axles to run through. I rather like the option of cutting those slots out and then bolting chunks of flat gunmetal in place. They are then drilled out to accept axles. This gives us the option of replacing the bearing surfaces if they wear out many years hence. But, in truth, down in these small scales, plain axles in plain holes seem to go on forever. The axles of my crude first steam engine, built more than fifty years ago, are still behaving themselves in simple holes in the frames. It may not be precision engineering, but it works.

MAKING THE FRAMES

The frames are marked out on 1.7mm (approx. 70 thou or 1/16in) mild steel. Wirral has that slight complication of having additional outside frames, so there are two pairs of them to be produced. I like to saw out blanks roughly and then bolt them together as pairs. As mentioned earlier in the book, I also add an extra strip of metal to the mainframes, to act as a jig for drilling out the coupling rods later.

The blanks are sawn and filed out. Drilling the holes comes next. The individual parts are separated. Any slight burrs are smoothed off and then the job is given a couple of coats of metal primer spray. Buffer beams are marked out from 3/4 × 3/4in steel angle, then sawn and drilled. There are also some corner pieces to make from brass or mild steel angle. The basic frames are then assembled. There is now something tangible to look at.

It isn't critical when the buffers are made. In the bigger picture that is a working garden railway, feel free to be happy with solid buffers. But if you have a desire to make them sprung, then by all means do so. As the photographs show, the driving wheels on my example are sprung. Before assembling the frames, I had riveted some horns in place. Mine were just sawn from metal, but there are plenty of castings to be had. The axle boxes were machined to shape in the lathe. In times past, I would either file these by hand or solder a bit of bar to a piece of plate to get the same end result. I am not very keen on hard work, but it has its place at times.

Because the leading wheels fit inside outer frames, there is an opportunity to give the axle a little end float in the bearings or axle boxes. This gives our loco the ability to go round sharper curves than

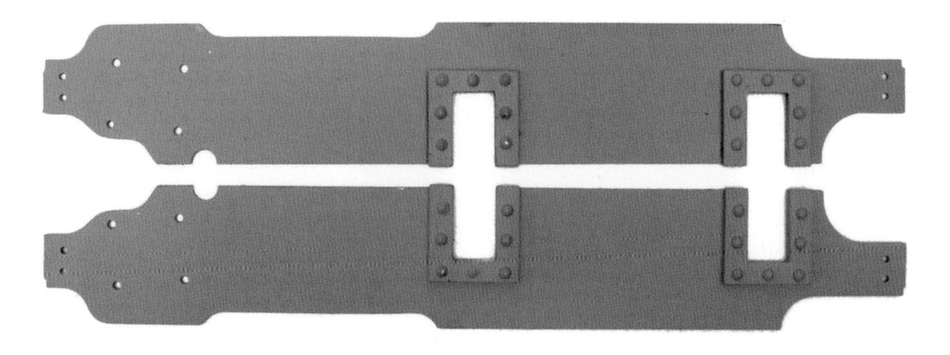

Mainframes with slots cut for axle boxes. Instead of hornblock castings, simple steel reinforcing pieces have been riveted in place.

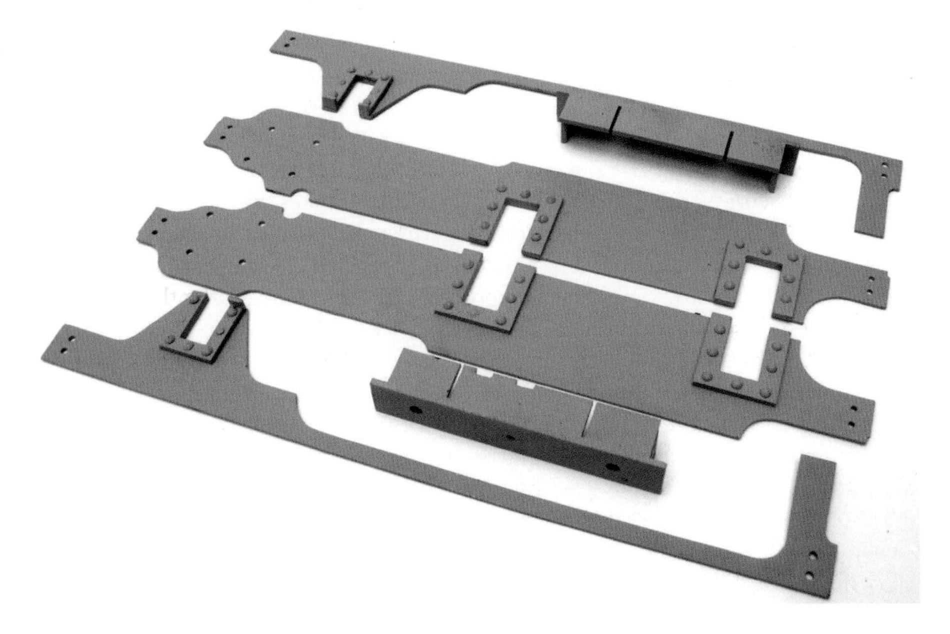

Inner frames, outer frames and buffer beams.

would otherwise be the case. Remember that, in 2.5in gauge, we are really looking at a minimum of 8ft radius (other than for very small 0-4-0 tank engines). In any case, small radius curves can look rather toy-like.

THE BOILER AND SMOKEBOX

This was made from a piece of 55mm diameter copper tube, 7½in long. Yes, you should be getting used to the mixture of metric and imperial sizes by now. It seems to go with the hobby, as you are discovering. To make the smokebox, a further piece of that tube, 45mm long, is sawn lengthways so that it will spring open a bit and slide over the boiler tube. The smokebox front is a plain disc of copper of 55mm diameter. Because the smokebox has been opened up a bit, there will be a gap. Cut another 45mm, then cut out a segment of this just large enough to fill the gap. The soldered joins will be hidden underneath the smokebox where it sits on the saddle – only you will know about them.

The boiler front and back will need discs cutting that fit snugly in the tube ends. This particular model was to be gas fired, so a single fire tube – 22mm copper tube – will run from end to end. Holes will have to be cut in the end discs for this. Lightly mark a centre line that will form the top of the boiler. Mark where the dome and the rear steam take-off point will go. Drill for these. Make the bushes, then silver solder them in place. This allows

The gas-fired boiler and smokebox.

you to clean out any swarf or gunge before making the boiler itself. Silver solder one end disc in place. If the fire tube has any cross water legs or solid rods (a 'porcupine'), insert the tube from the opposite end and add the other end disc. Both ends can now be silvered soldered up, attaching the discs to the main tube and also fixing the fire tube in place at the same time. Place the completed boiler in a pickle bath for a day and then rinse out thoroughly.

The smokebox can be drilled underneath for the exhaust pipe from the cylinder, as well as the main hole for the chimney. Add the dummy smokebox door detail. Don't use a white-metal casting because there is a big roaring jet firing a lot of heat towards it. You may be able to get a gunmetal casting, however. One useful little dodge is the humble tin lapel badge. Remove the spring and peel back whatever label it has clamped round it – and you have nice little smokebox door shape, often in plated brass. They come in all sorts of diameters, so they can make the basis of a smokebox door assembly. But, really, you will have reached a stage now where you could make your own by cutting a disc in thick brass, profiling it in the lathe, and then adding straps and dummy hinges by silver soldering them in place.

FITTINGS

A properly working safety valve is absolutely essential. Whilst making one is now within our capabilities, I am going to suggest that you buy a ready-made one for your first engine. It takes all the guesswork and worry out of getting the boiler tested later on. If you want a pressure gauge or a water gauge then buy them in too. You can make a needle valve for a regulator, but, again, it is easier to buy a ready-made item. An ordinary model globe valve – as sold for stationary engines – is perfectly alright (an inline lubricator is something we can easily make ourselves), but in this catalogue of things to buy in, I am going to suggest that we include a ready-made gas tank assembly and jet. Apart from a pressure gauge, which is difficult for most people to make accurately, we can graduate to fabricating all of these fittings in later models. For now, though, opting for the simple life has much to commend it.

Chimneys and domes have been discussed previously. In the case of this engine, I turned the chimney saddle out of copper and then annealed it, before hammering it down onto a slightly undersized piece of metal bar. Two blows was all it took to shape it. I didn't flange the bottom of the dome, reasoning that this might be a bit tricky for a beginner. However, it is bored inside to be a sliding fit over the safety valve (which also doubles as the boiler filler, when unscrewed). I had put that little turned pimple on top of the dome, but you can be content with just drilling a small hole. I also simply screwed the regulator into a bush in the back of the boiler. If I had wanted a pressure gauge (and/or a whistle), I would have made a steam fountain/manifold.

The chimney saddle has been annealed and then beaten over a round bar.

The completed chimney.

At this stage, test the boiler. Make a little wooden cradle to hold it. If you have a commercial safety valve, which is known to blow off typically at 40psi, and a pressure gauge, you can keep the regulator closed whilst having the burner running. Don't expect the gauge to read exactly 40lb when the valve lifts; such things are approximate. But you will be looking for any little weeps at your soldered joints. If you have built the boiler according to the words and music, this is unlikely.

If all has gone well, the next step is to enquire with your particular gauge association about getting your boiler properly tested if required. This process will hydraulically pressurize the boiler to a higher pressure (typically twice working pressure), measured against an accurate master gauge. If there are no problems, you will be issued with a certificate, which organizers of public exhibitions may ask you to produce. In any case, it is good for your peace of mind to know that your boiler is safe. If you buy a ready-made gas tank/regulator/burner assembly, it may or may not come with a certificate that says that the tank has been tested to approved standards. This seems to be a grey area (*see* Appendix III). Incidentally, if the pipe to the burner is in the form of a heat-resistant plastic tube, look for a suitable long spring that slips over it neatly, and forms a little bit of armouring.

At this stage, I suggest that you clean and dry your boiler thoroughly and then paint it. Do a thorough job of painting and then leave it to harden off whilst you get on with other things.

With the boiler and smokebox made, they can be propped in place to establish the centre height above the running board. In the case of Wirral, this was 40mm. The smokebox saddle can now be made, as well as a rear-frame stay that has a semi-circular notch cut in it to suit.

PAINTING INSIDE FRAMES

It was quite common for the insides of frames to be painted red. If you do this in the model it adds a certain something – a bit like a glimpse of stocking. But I suggest that this is not suitable for spirit-fired pot boilers, where that nice paint surface would quickly get ruined. In that case, stick to heat-resistant black.

AXLES AND WHEELS

The turning of wheels and axles has already been described in detail. We looked at a jig for drilling

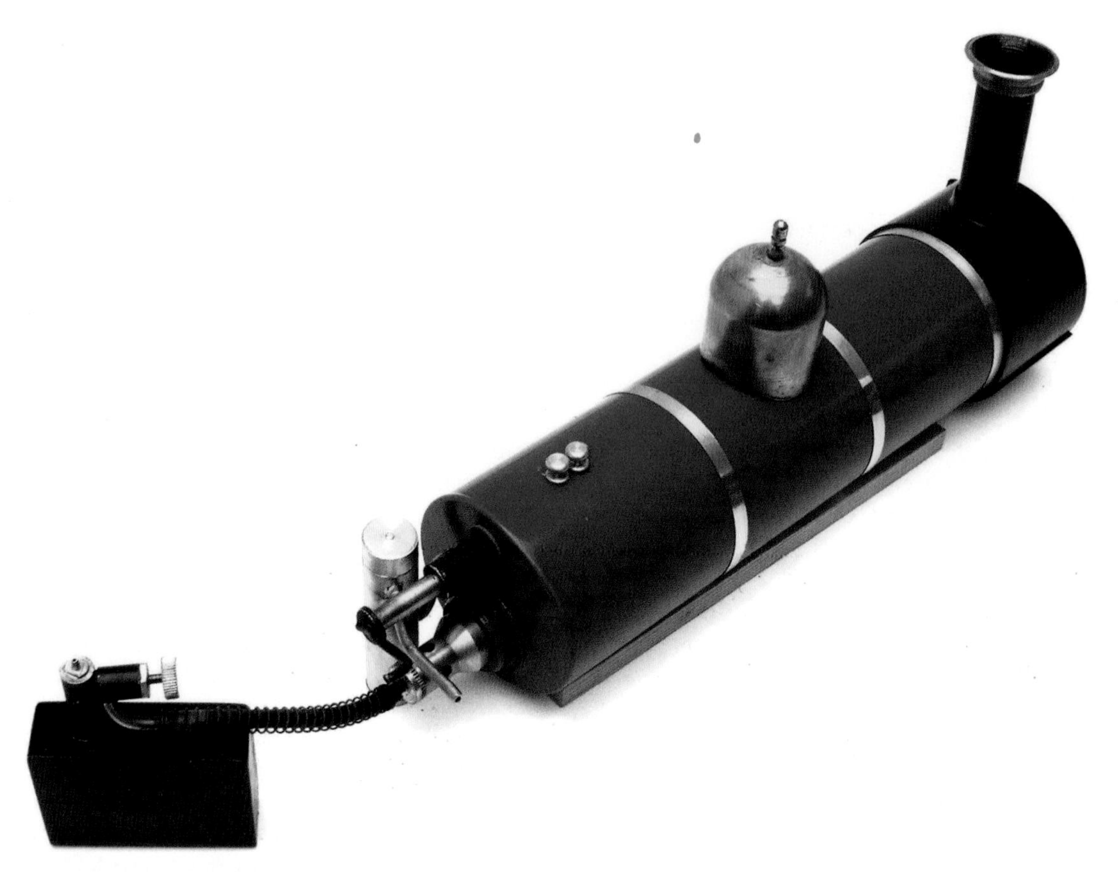

The completed boiler, with gas tank and jet, lubricator, regulator and boiler bands (note the spring that wraps around the gas delivery tube).

The cylinder of the prototype was fabricated from steel bar. Gunmetal would have been better.

crankpins and how to fabricate coupling rods. There is nothing special about those in this model. There are no bogies or pony wheels to think about.

CYLINDER ASSEMBLY

The cylinder and valve chest were machined from the solid. Mild steel was used for the prototype of this model, but it would be better to use phosphor-bronze or gunmetal. The cylinder block is 32mm wide, 28mm high and 45mm long, which was achieved by facing the original cob of metal off in the four-jaw. This was then bored through and reamed to ⅞in diameter (or approx. 20mm). Surprisingly, it won't matter if the bore is a millimetre larger or smaller – provided it is accurately reamed and the piston is machined to size accordingly. The purists may be horrified at this apparently casual approach, but, at this stage, the aim is to produce a beginner's simple working engine, rather than one that wins model engineering competitions. If you are working from an established set of drawings, then, of course, stick to the dimensions specified.

The internal passages were drilled through and the ports milled. There are two different cylinder covers to be made. The one at the front end is just a plain disc, with a small 'step' in it to recess slightly into the cylinder bore. At the rear end there is also a gland to be shaped from solid metal. I made this in two parts and silver soldered them together. Even if you only have one cylinder in a locomotive, it is worth making a cylinder cover drilling jig, perhaps using a suitable washer. Ideally, the holes want to be equally spaced and on a true circle. You know you have achieved this if the cover will screw onto the block in any position. But if you are slightly out and the cover will only go on in one particular location of holes . . . well, it's hidden nicely down between the frames. The jig will not only allow the holes in the covers and the block to line up, but they will also help in cutting and marking out the gaskets. In small engines like these, oiled brown paper will be perfectly acceptable as gasket material.

VALVE CHEST

This is also 32mm high but is only 16mm wide. These dimensions, for 2.5in gauge locos, are quite critical. Together with a valve chest cover of 2.5mm thickness, we need to make the total width of the assembly 51mm wide, so as to fill the distance between the insides of the frames. If there is any cumulative error in your machining, don't panic – you can make it good by using a chest cover of a different thickness.

After shaping the metal block to size, I'm afraid there is a bit of hard work ahead, to cut out the hollow centre. Mark out the size of this and then drill the largest hole you can in the centre, without

The connecting rod and eccentric rod being roughed out.

crossing the marked lines. (If you have milling facilities, it makes sense to take advantage of them.) Then settle down with a square file (it helps if it is fairly new!) and patiently file in all four directions out to the boundaries. This is yet another of those jobs where I don't try to do it all in one go. I leave the job set up in the vice and then repeatedly give it five minutes of file strokes between doing other things. Of course, if you are working from castings this inner space will already be cast in for you. There will be holes to be drilled in the cylinder and valve chest to admit rods through glands. The valve chest cover will need to be secured in place by countersunk screws so that the entire assembly will fit neatly between the frames. It will be held in place by some studs (headless screwed bolts). I suggest making a small drilling jig from a piece of mild steel in order to ensure that the holes in the frames line up with the holes in the cylinder assembly.

There will be a gasket to be made to fit between the cylinder block and valve chest, and another between the valve chest and its cover. That cover might well be one of the frames in the case of an engine like this.

ASSEMBLING THE FRAMES

The frames and the buffer beams are screwed together, sandwiching the cylinder assembly between them. Lightly tighten things up. Lay the frames on a truly flat surface, such as a piece of plate glass, and check for any wobbles. If there are any, slightly loosen the screws and give a twist with your bare hands. It should only take a minute or so to ensure truth. In the unlikely event that your measurements have been a bit out, the front and back buffer beam might not be at a true right angle. This is rare and unfortunate, but it is not critical. What *is* important is that the holes for the axles are exactly opposite each other, meaning that the axles will be at a true right angle to the frames. To check this, use a couple of long *straight* rods (12in or more) of the same diameter as the axles. Put them through the axle holes. Carefully measure the distance between the ends of the axles on both sides. The measurements should be the same. In the case of Wirral, there are outside frames to be bolted on as well.

FITTING THE WHEELS AND SETTING THE TIMING

This is a good time to offer up the axles and wheels. The aim is a free-rolling chassis. This means that the driving wheels need to be quartered. Make and fit the coupling rods. This means turning the crankpins to shape. Some people like to screw them in, but I find that a tight press fit is quite adequate (using a vice with protected jaws as a press tool). The chassis should roll freely with no little stiff spots. Be prepared to adjust things until this is achieved. When this happy state has been arrived at, I like to put a drop of superglue at the joint between the axles and wheels. This temporarily holds things in place whilst I drill a small hole (say, a tapping size for 8BA) halfway through the inside of the axle hole in the wheel and the axle itself. This is then threaded and a screw inserted, with a drop of Studlock high-strength adhesive. Cut off the

The crank axle; it still needs cutting to length.

The cylinder and motion have been installed between the frames and paint has been applied, whilst it is easy to get at.

protruding part of the screw and smooth off neatly with the wheel.

The connecting rod from the cylinder is now connected to the crank of the driving axle. If this is slightly tight, just give a touch of fine emery paper to the crank itself. If it is a bit loose, smooth a little off the faces of the split big end – slightly tight at this stage is what we are ideally looking for. The apparent length of the rod can be slightly adjusted so that the

A simple jig for fabricating coupling rods. The 'brasses' are held in pins that sit in holes that were drilled out when the frames were being made.

piston travels from end to end of the bore, without quite banging into the cylinder covers. The eccentric strap from the valve rod is bolted over the eccentric cam now, providing us with a slight dilemma. The normal procedure for setting valve travel is to take the valve chest cover off and then adjust things so that the valve itself just uncovers a crack of the steam ports on both sides. However, our cylinder assembly is trapped between the frames. In this case, we will resort to compressed air. The ideal thing is a little paint spray compressor. Next best is a friend who has one! In days of old, we would get a friend to use a car foot pump (after having made a small adaptor for a Schraeder valve) whilst we set the timing.

There are two elements in setting the timing. The first is adjusting the position of the valve on its rod, as described above. By screwing the valve on its rod, we can adjust the position of its travel across the port face, but we also need to set the timing of this relative to the stroke of the piston. This is achieved by rotating the eccentric on the axle and then locking it in place. A good starting point is to set things so that the valve just starts to move when the piston rod is about halfway along its stroke, and does so running in either direction.

Having got over the sheer delight of seeing it actually working on air – always a landmark moment – it is time to tweak things. Adjust the rotation of the valve eccentric so that the performance seems the same in both directions. Next, adjust the position of the valve on its spindle by screwing the latter in or out of its bush on the valve rod. Keep adjusting until you hear an evenly spaced 'pah-pah-pah-pah' from the exhaust, in both directions of travel. You will have set the timing entirely by ear – quite an achievement, really. Your reward will be connecting the boiler temporarily to the engine and firing it up. And this really is one of life's good moments, when your first engine, propped up on blocks, is ticking over in steam.

PLATEWORK

Screw or rivet some angle (actually, I used ¼in square steel rod) to the inside of the outer frames, then cut and fix some running boards down in place. Typically, these would be in 1/16in mild steel. If you were building a conventional locomotive, without those outer frames, then the procedure would be to fix the angle to the running boards first. These are the valances. Quite possibly, there will be small downward extensions of these at each end in order to blend into the buffer beams. Make these separately, silver soldering them in place. Also needed are two infill pieces at the front and back of the loco, covering the area between the frames, so as to present a flat surface all the way round. Cut out in cardboard first, by trial and error. When satisfied, cut the shapes from metal sheet.

A support for the boiler has been fitted. You will also note a proper wood cab floor now in place . . .

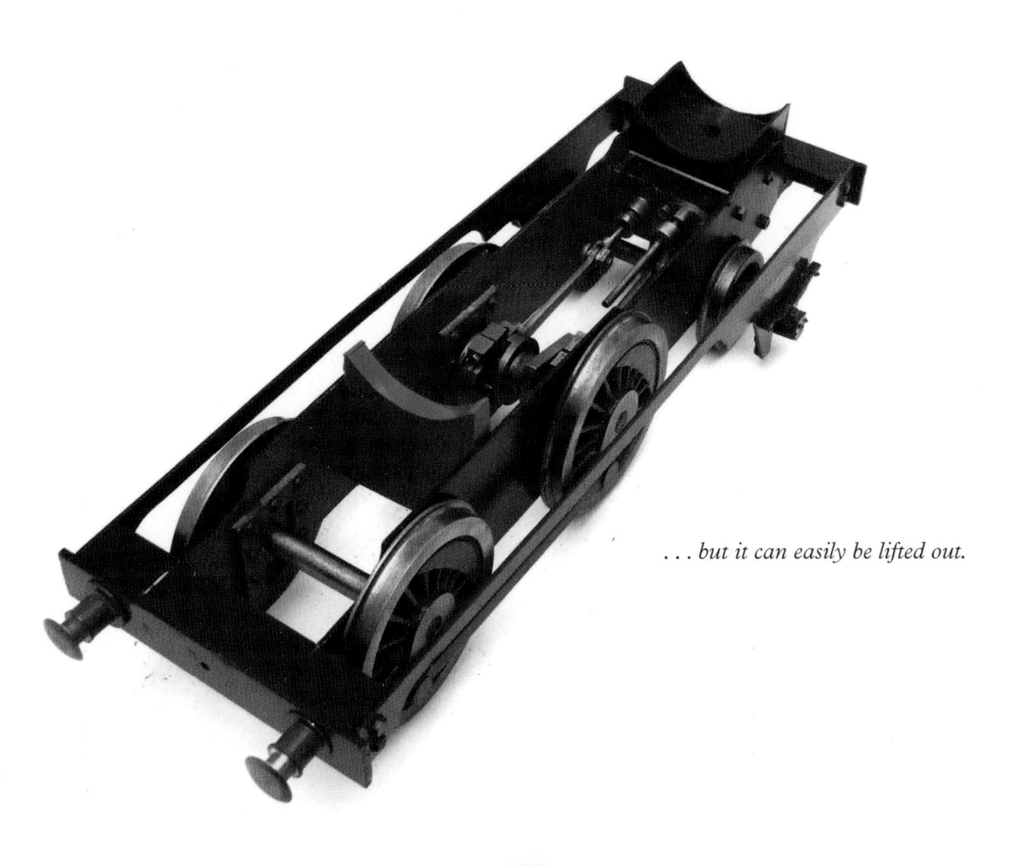

. . . but it can easily be lifted out.

Making side tanks and bunkers will be familiar ground by now.

The side tanks and rear bunker might well be in $^{1}/_{16}$in brass. Now if you could get that material in long strips 2$^{1}/_{2}$in wide, what a labour-saver that would be. This will shock the purists, so let this be our little secret – when considering a loco design, try to organize platework to use existing set widths of material. And if this means that the design is just a tiny bit out, you may be able to live with that, especially for your first efforts. It is worth it to have those lovely straight and parallel lines.

We dealt with assembling a basic brass box shape in Dempsey (*see* Chapter 4). Much of loco-building is just that – but with different dimensions. We screw, rivet or soft solder pieces of rectangular sheet to short lengths of angle to form the shapes we want. I am still addicted to using superglue to hold pieces together temporarily whilst they are being more firmly fixed. There may not be a need for a full assembly jig this time, but a useful device consists of two pieces of wood, several inches high, that are screwed down to a baseboard so as to form an accurate internal right angle. We can push corners of metal against this whilst they are being worked on.

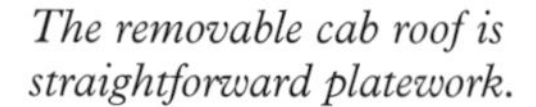

The removable cab roof is straightforward platework.

Suddenly a steam test is with us one bright sunny morning.

Case Notes:

Terrier

Let us take an example of a very specific and well-known prototype and run through the thought processes needed to convert it into a sensible working model, in the manner of Wirral. The Stroudley A1 class were tiny engines. Actually, we can use the word 'are' instead of 'were', because many of the rebuilds to A1X form are with us still. I was fortunate enough to be a child within earshot of Fratton engine shed, Portsmouth, and saw the Hayling Island locomotives on a daily basis. So I still have a soft spot for them.

I will suggest that their smallness would make it difficult for a beginner to build a live steam example in O gauge. It would not be impossible for someone with determination, but I estimate that an eight- to ten-minute run would be as much as could be hoped for at any one time. Even in Gauge 1 it is a small piece of machinery, though it is certainly more viable. In 2.5in gauge it makes a pretty little model and my thoughts will be mostly directed towards that.

The 0-6-0 chassis with a single cylinder between the frames would be very much along the lines of Dempsey, described in Chapter 4. If the engine was intended for garden use – particularly down at low level with hazardous twigs and snails to contend with – I would leave off the sandpipes and probably the brake gear as well. I take the pragmatic view that better a simple reliable engine, giving good service, than a Gold Medal winner constantly being repaired.

I would also be inclined to make a stout metal frame to go around the inside of the top of the cab, onto which would drop a removable cab roof, for easy access. To keep a neat outline and to preserve a good paint job, I would plump for gas firing. If the gods were with me, I might be able to buy some etched brass body components ready made (sometimes referred to as 'scratch-aid'), but I would examine these carefully to see if any needed discreet reinforcing to cope with the life of a hard-working steam engine. Certainly I would suggest making the frame's connecting rods and buffer beams in mild steel. Incidentally, that cab roof (the shape was intended to prevent a resonant drumming noise) is awkward to make. In 5in gauge, I beat one in thin sheet brass over a wooden former. But for these small scales I can only suggest filing and patience.

Terrier details came in a wide array of permutations, changing constantly. The drawing shows one in Billinton livery with (worn-down) wooden brake blocks and before the rebuild to A1X, with an extended smokebox. If you are new to Terriers – or indeed any other prototype – then it is a case of doing some homework first. I always found this a pleasure rather than a chore.

The generous dome can conceal a safety valve cum filler, but a commercial boiler might have bushes already fitted at the back end and would thus live inside the cab. There is plenty of room for a gas tank in the

Terrier *continued*

bunker and a displacement lubricator can fit in the cab, with a drain cock coming out beneath the running boards. As usual, a pressure gauge can peek sideways out of the cab, facing the side you will see most as it works on your layout. An alternative is to have it visible through one of the round spectacle openings.

Perhaps you will allow me the indulgence of suggesting another prototype from my past. I had a hand in rescuing the Fox Walker 'Margaret' many years ago. She was built for the North Pembrokeshire Railway, got absorbed and rebuilt by the GWR, then eventually declined into industrial use and was finally put in a shed and forgotten about for decades. She now resides, vaguely cosmetically restored, at Scolton Manor in Pembrokeshire. She is unusual in having no balance weights on the wheels. I have a soft spot for her, but she also represents the vast numbers of saddle tanks that humbly worked around the country. Incidentally, I find that the easiest way to make ends for semicircular tanks is to cut out a complete ring and then saw it into two segments.

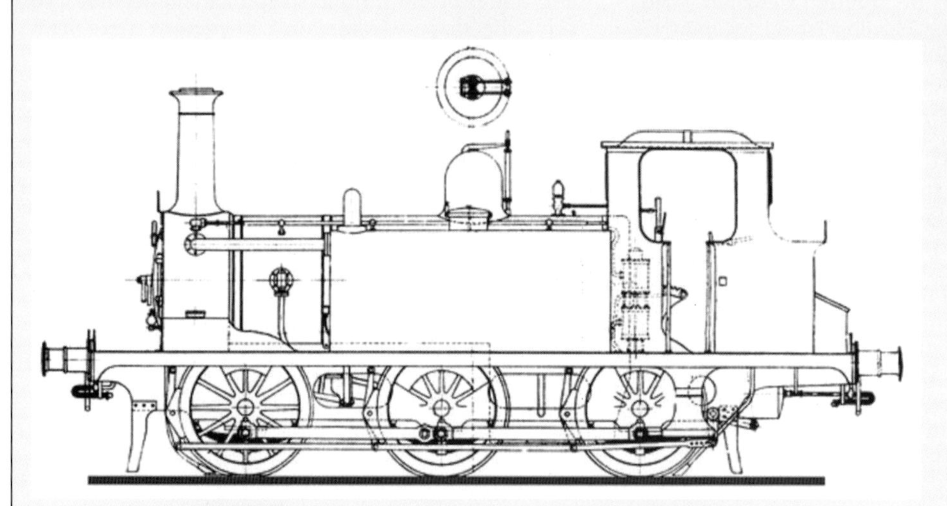

The original A1 class had the shorter smokebox.

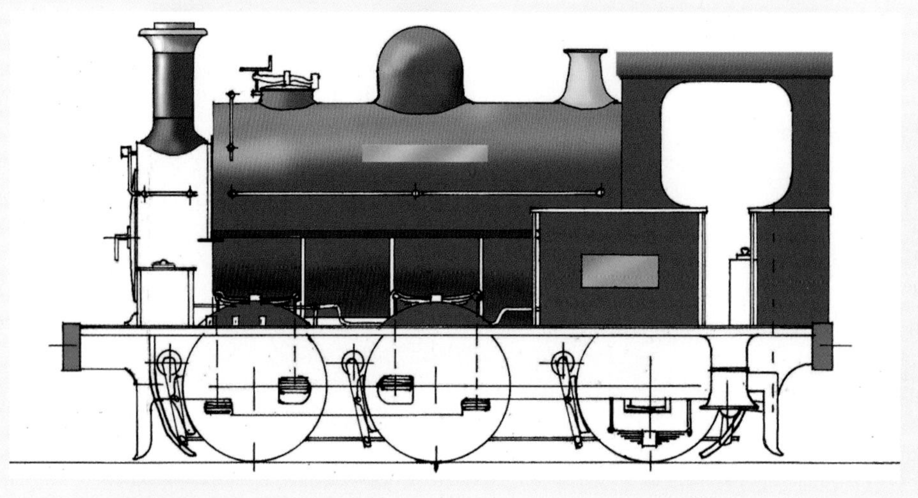

Design for 'Margaret', an old flame of mine.

Terrier *continued*

'Margaret' as she actually is. Like the Terrier, she could be another 2.5in gauge locomotive like Wirral, merely with differently shaped components.

Don't forget to think out how the parts fit together. For example, the ends of a tank may fit inside the long sides, which means that we have to allow for the extra two thicknesses of metal that would otherwise increase the overall width. Add any details at this stage, for example tank fillers or coal bunker doors. When these complete 'boxes' are finished they can be drilled for screwing up from underneath the running board. Don't be afraid to use self-tapping screws, but don't fix things permanently for now. They will be much easier to spray paint prior to assembly.

As with Dempsey, it is a good idea to soft solder any number or name plates into position before spraying. The entire side, including the plate, can be sprayed body colour. When it is dry, overpaint the plate by hand in its proper background shade. When this is hard, a scalpel blade, lightly applied, will reveal the raised polished metal. Give a final touch over with a piece of the finest emery paper, or even a hard rubber, just to give it a final polish.

I am rather partial to putting a planked wood floor in the cab. It is one of those easy jobs that gives a model an instant lift. I usually employ a thin piece of ply on which I have either scribed lines or glued thin wood planks. Usually they would become grey and grubby very quickly, but I like to stain them a moderately dark colour and then apply a couple of coats of satin varnish.

The boiler sits down on the front saddle and the rear stay, but is only lightly restrained. Strictly speaking, the boiler should be free to expand slightly along its length as it gets hot, so it only needs to be held down to the saddle with a couple of screws. This may seem 'unengineering-like' but it works. In my Wirral, the gas tank merely sat in the cab area. It could be located inside the coal bunker if you prefer a cleaner appearance.

The engine now presents as a nearly finished open cab engine that has very accessible controls. A common ploy is to fix a cab in place and have an easily removable or hinged roof. However, this design cried out for the cab to be instantly removable, so it was made up as a solid unit, soldered up in brass.

FINISHING OFF

All of a sudden, we find ourselves approaching completion. Things seemed slow going at times,

One last look at Wirral: a pleasant little ragamuffin.

but the end comes suddenly. There is a temptation to rush the last bits, just to see it running. I promise you: I know the feeling well. For a practical working model, it may be better not to dress things up with too much fine detail that can get damaged. I had put a compressor and a receiver tank on mine, because it was closer to the prototype I had seen. I should have fitted brake pipes on the buffer beams. There are no cab steps or handrails.

The boiler bands are merely brass strip cut to length, with right angles turned at the ends and 10BA nuts and bolts to pull them tight. If you look at the side drawing of Dempsey, you can see this clearly. Again, the easiest thing to do is cut a dummy one from thin cardboard first, to get exactly the right length. I suggest leaving the cab windows unglazed. When everything is painted and assembled, add a dummy coal load in the bunker.

Because Wirral is a large shiny engine, even more than other locos, it is worth pampering her to keep the nice appearance. Make a bit of dummy track – even with wood rails, if you wish. Get into the habit of lifting this base, whilst cushioning the buffers with your thumbs. This is much better than getting grubby fingerprints on her. It also cuts down on the tendency of paint to wear away on edges that are constantly being handled. Make up a wooden carrying case that the engine and base can slide into. Pack it with soft cushioning material. An alternative is to make a couple of inverted J-hooks that bolt into recesses in the wood base. These hook around the rear and front axles and are lightly tightened up with recessed nuts underneath.

As was mentioned before, I like to see an exhaust pipe that allows the steam to jet upwards freely. I believe it gives a freer exhaust. But it does mean that, when firing up, great jets of hot water shoot up from the chimney for a few moments. As noted, a right-angled plumbers' bend will divert this sideways. However, this is such a pretty engine that you might like to consider cutting a piece of plywood that has a hole cut in it, through which the chimney pokes upwards. It provides a little umbrella for our engine until it is steaming cleanly. At the end of the run give the engine a whiff of WD-40 and a wipe-down with a soft rag, whilst it is still warm.

The job is done: and very satisfactory it is too.

CHAPTER 10

Motor Tank

Thus far, we have used various stratagems to make life a little easier for ourselves, but now we will tackle a detailed design for a scratch-built engine. What we have here is a fairly close scale, Gauge 1, model of an LNWR 2-4-2 Motor Tank. It is not super-detailed, but will be a good, honest runner. In a long career, the noted loco model-maker Harold Denyer has designed and built a large number of locomotives to this formula, so you can be sure that this engine will be an attractive and reliable first entry into the world of traditional small-scale engineering. The drawings are typical of those that you might have encountered over many decades. Count the fact that they are his original

hand-drawn artwork as a bonus. To some of us, there is something aesthetically pleasing about this. It speaks of a pre-digital age, now rapidly disappearing.

As mentioned throughout this book, you will see that there is a total freedom in jumping from one set of measurements to another. There will be fractions of an inch, metric dimensions, sheet metal described in swg (standard wire gauge) numbers, BA sizes and letter/number drills. It seems illogical, but there were always good reasons. Some measurements seem to fit comfortably with our needs. Details of wheel castings can be found via the Gauge One Model Railway Association.

> 'A table of cycle threads may now be long obsolete – but there are still plenty of old bicycles around.'

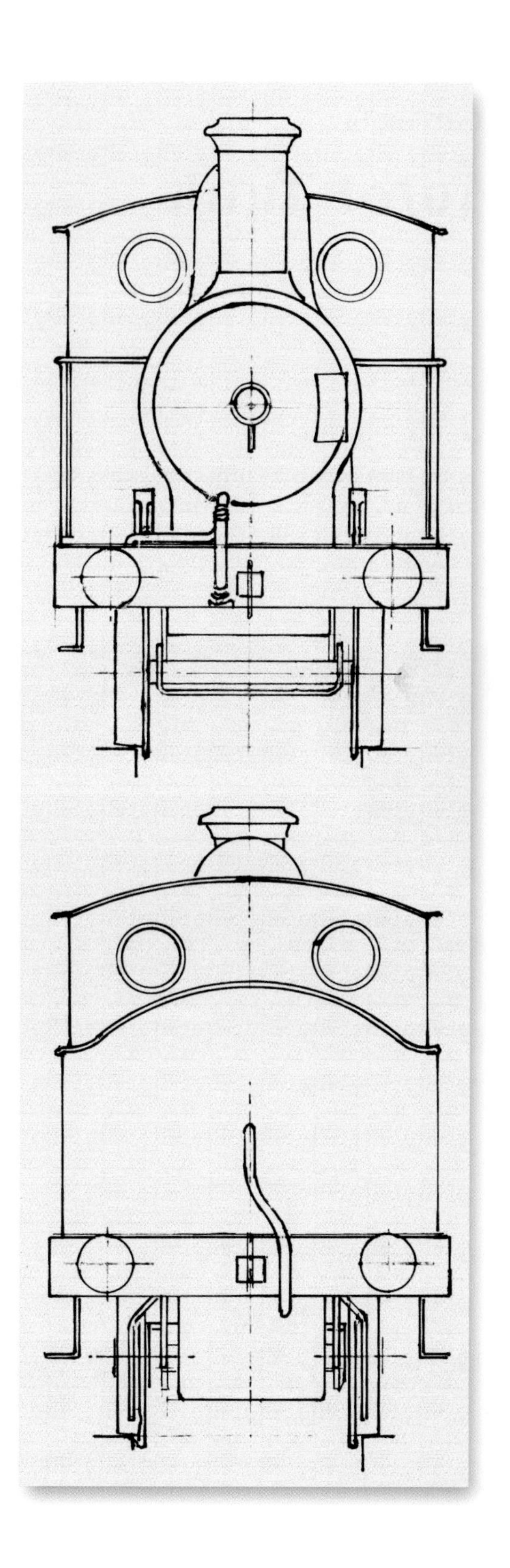

As you progress further down this road you will become comfortable with this variety of tables. Don't try to learn them: just absorb them as you go along. There is a movement towards total metrication. This could make life much easier and, if you want to go this route, I wish you well. But it would exclude a great heritage of loco drawing and building that you might want to draw upon.

The prototype was chosen because it has very simple platework but is to an attractive outline. The loco will go round comparatively sharp curves by Gauge 1 standards. There is a single cylinder between the frames, operated by slip eccentric valve gear, similar to Wirral. The boiler is made from 1¾in diameter copper tube. It features a firebox, open at the bottom, which will host three wick tubes. The heat is drawn into a single fire tube, which also has cross water tubes. This combines to give something that will steam like a witch. Those smaller diameter wheels can also contribute to better slow running.

The spirit is contained in a traditional chicken-hopper pattern of tank. This will give a long duration of run. An adaptor for an Enots filler is shown on the plans and so the boiler can be constantly topped up with water whilst under pressure. A Goodall valve could be fitted and it would be possible to fit a small hand pump in a side tank. Both tanks would contain water, with a balance pipe

THE PROTOTYPE

The London and North Western Railway built a large number of moderate-size tank engines for a variety of everyday uses. The 2-4-2 tank engine was particularly suitable for running comfortably in both directions whilst doing a job of work. One variation was a locomotive that had smaller diameter driving wheels for fast acceleration between stations. A number of these were fitted for motor-working. A trailing coach would have a driving compartment at one end, from where the engine could be controlled. This required additional plumbing to coach and engine, but it offered fully reversible running in the same way as a modern multiple unit electric or diesel train.

The story of LNWR tank engine details is extremely complicated. Smokebox door handles, coal rails, pumps and many other things were fitted and removed at different dates to different members of a particular class. For example, on a detailed model, there would be a choice of two different types of lamp bracket at different times. I recommend that you join the excellent LNWR Society should you want to take your research further.

Overhead view of the 2-4-2t. This example is fitted with coal rails on the bunker.

leading between the two, under the boiler. A 'dead-leg' displacement lubricator is located inside the left-hand dummy side tank. With this design, we also reach a new landmark. Lots of dimensions are given on the drawings, but the stage has now been reached where the builder has to think for himself. There is a need to work out ways to do specific tasks. If you are on your own – without someone who is experienced looking over your shoulder – and this is your first engine, then enjoy working out your strategies; it is part of the pleasure.

FRAMES

The original specification is for 18swg mild steel. The frames will be formed out of strip that is 40mm wide minimum. The loco is 299mm long, but you will need to add a little more to take account of Harold's liking for bending angles. This is an alternative to using small angle section for joining components at right angles. If you are a newcomer and would like to bend angles, I have a few thoughts that may be worth sharing with you. Firstly, make some smooth jaw liners for your vice, perhaps made from a couple of offcuts of thin aluminium sheet. Practise bending angles with some offcuts first. Don't try to hammer angles over directly. A piece of square steel rod, held horizontally between the hammer and the steel strip, keeps the bend flatter and truer. Perhaps it would be useful here to repeat a tip we looked at in Chapter 4. Don't cut the strip exactly to length and then try to get the bend in exactly the right place. Make it a bit overlong, put the bend in and then cut the main part of the strip, and the little bent portion, to length.

Note that, in order to reinforce the frames at a weak point, the angle pieces that support the running plate are actually bent parts of the frames as well. You will need to allow for this when marking out the material for them. Here, I suggest that you cut a frame out of card, including the material for the folds. Try out the folds and also mark the locations of the holes. This lets you become completely familiar with what is going on before starting to cut the metal. The long top bend towards the rear of the frame gives real strength to the narrow neck. Again, I urge a beginner to have a practice bend on a piece of scrap steel sheet first. Catch this in the vice and hammer it over to form a true right angle. Don't be afraid to put a length of, say, ½in square mild steel lengthways against the sheet and hammer that down. This will avoid a wavy line. Similarly, if you do detect some slight unevenness, it can be hammered flat against the chunk of steel bar or even the vice jaws.

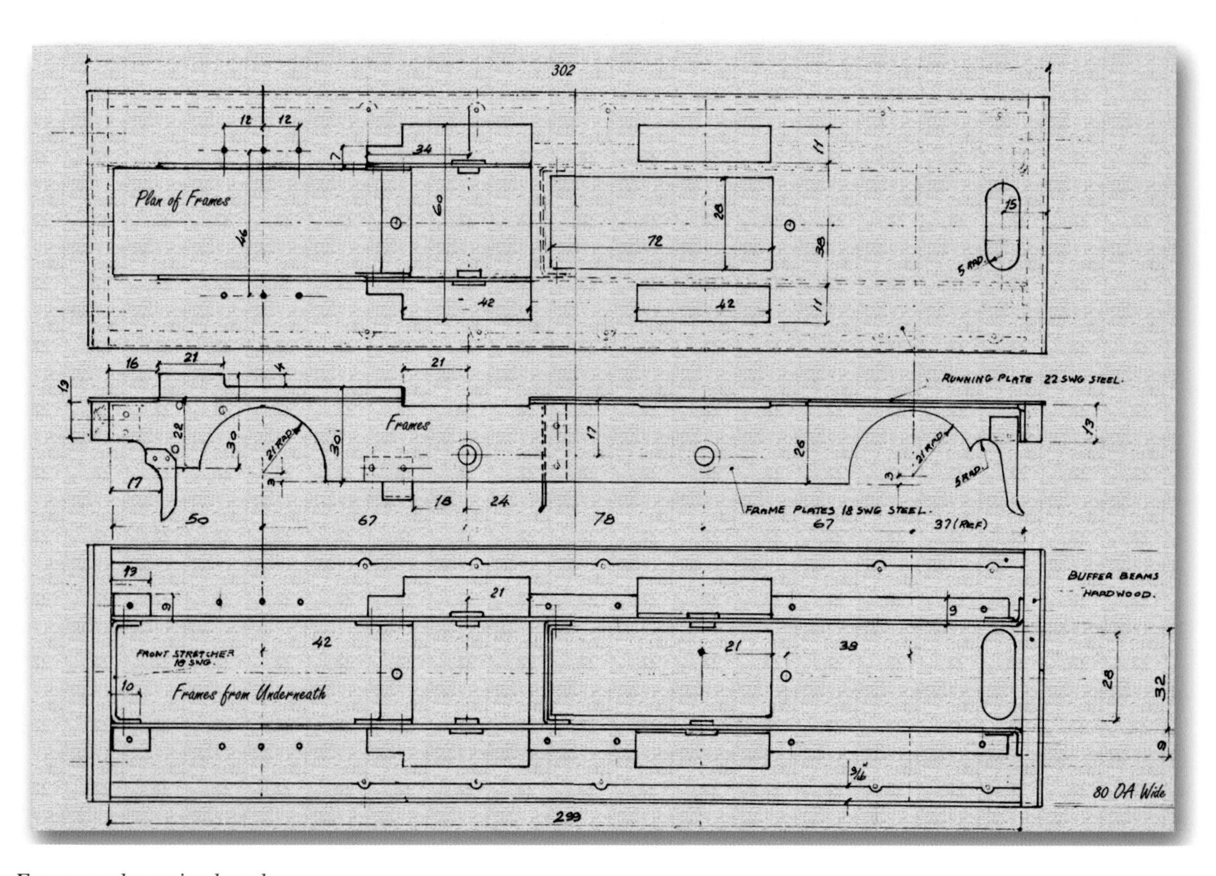

Frames and running boards.

Look for something with the right internal diameter to which to file the curved frame openings.

You will see that the mainframes are mostly straight lines except for those two circular cut-outs for the smaller wheels, which are 21mm diameter. If it were me, I would cut these out with a fretsaw, with a metal cutting blade – and cut them undersize. Then they could be finished to size with a half-round file. If you are really nervous about getting this right, think about turning up a doughnut shape in thicker steel, in the lathe. The inside of the doughnut is 42mm diameter. Grip the frames in the vice, along with the device and then you can file down to this inside curve – it will be perfectly accurate.

Holes are drilled for the axles. Mark the spacing with accurate centre pops, then drill pilot holes through both frames and an extra steel strip, which will be a jig for drilling the coupling rods. Again, tie a label to this and put it away safely. The axle holes in the frames can then be opened out to $^{3}/_{8}$in.

There are separate guard irons riveted to the frames at each end. You could cheat and merely cut out the profile of the frames to include them, allowing a little extra length to accommodate to the fact that they are bent outwards. The difference is hardly noticeable out in the garden. Or maybe you would prefer to make them separate – just because it is the right thing to do.

The frames are held together by several stretchers and the buffer beams. You will note that these beams are made of hardwood, 4 × 13 × 80mm. There is something satisfying about using the prototypical material. As with Dempsey's dumb buffers, you should be looking for a dense, close-grained hardwood. Old hardwood, if sound, is much to be preferred to strips of new material. A builder's merchant could well offer you a fairly poor, open-grained redwood that is still horribly unseasoned. I have a prejudice against ramin. It is dense-grained but splits at the slightest provocation. These wood beams are faced with mild steel.

THE RUNNING PLATE

The single large running plate is a complex shape to cut out, but it is all straight line work. The cut-out with rounded ends at the rear can be made by drilling two 10mm diameter holes and sawing away the space between them. Catch one end of a metal-cutting fretsaw blade in the frame and then thread it through one of the holes. Clamp the other end in the frame (if you were born with three hands, this makes life easier). Saw away the bulk of the material, finally finishing with a file to give straight edges that blend nicely into the end curves.

There is an option to make the running plate in separate pieces. There would be two long running boards that sit on and outside the line of the frames. A footplate between the frames would require additional angle inside to which to fix it. Finally, there is the small infill piece over the front buffer beam area. This approach also has an advantage for the beginner inasmuch as the valances (angled pieces to add rigidity to the outer sides of the running boards) can be attached more easily if they are fitted to the running boards away from the loco itself. You will note that there are small notches in the valances to clear screws that will later hold down the tanks and so on.

ASSEMBLING THE FRAMES

It has seemed slow going so far – there has been a lot of careful sawing, filing, drilling and bending with little to show for it. But with the valances

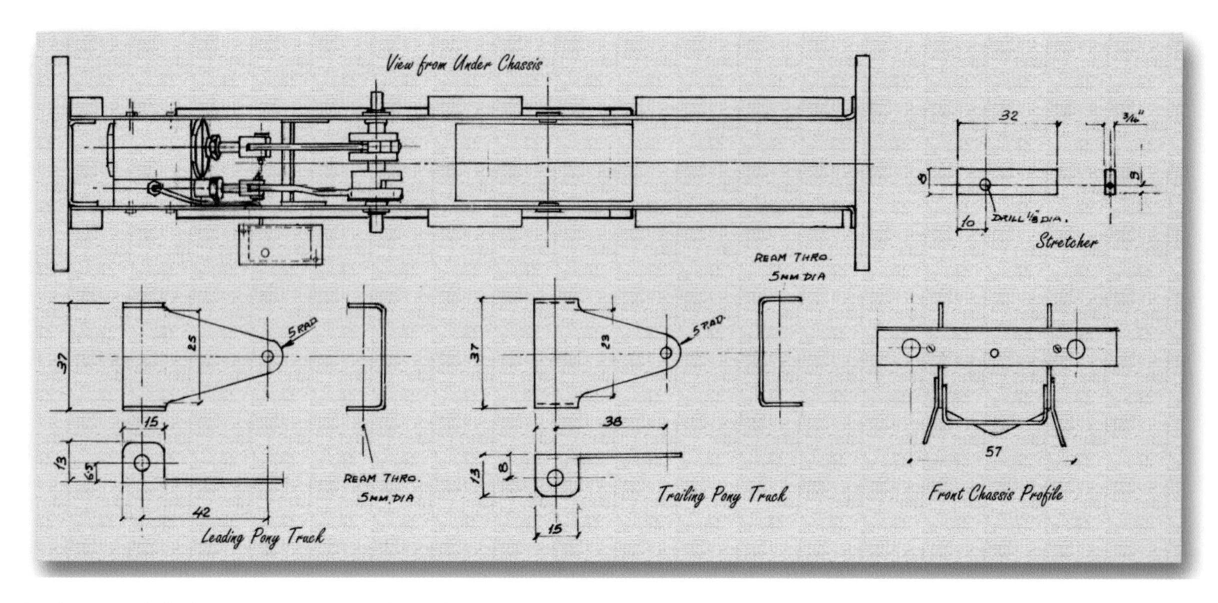

Underview of the frames plus pony trucks and so on.

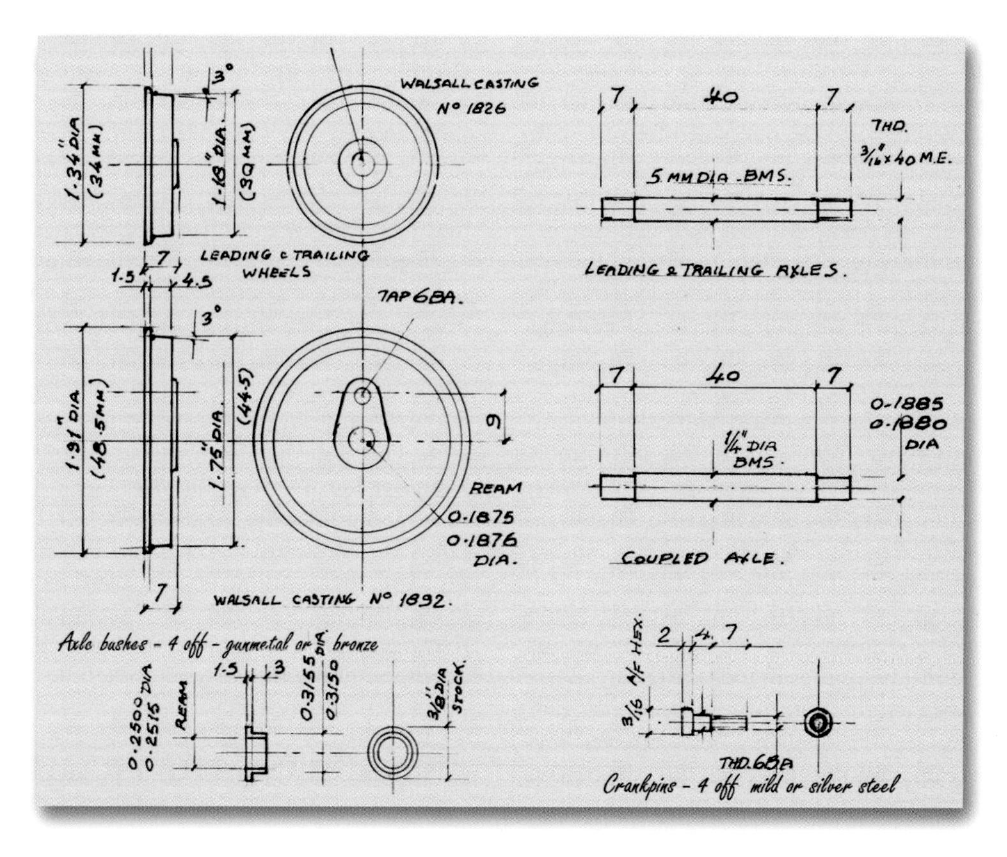

Wheels, axles and bushes. Note that the dimensions of the bushes are different to those in the text, which describe a beginner's way of making them.

fixed to the running boards and with rows of small snap-head rivets (in neat straight lines please), everything can now be bolted up and you suddenly have a completed bare chassis in front of you. This is most encouraging. Again, we will follow the by-now familiar practice of not bolting things up too tightly at first. It is time to check for squareness all round. It is *vital* that the axles are at a true right angle to the frames. Put some suitable rod through the axle holes, stretching out either side. Use a small metal square to check for truth. Also, lay the chassis down on a flat surface that is just short enough to allow the guard irons to overhang at each end. The chassis should sit flat on these, without any rocking. Having proved that everything is square and true, the frame assembly can be put to one side for now, safe in the knowledge that it will have to be dismantled again later on!

CYLINDER ASSEMBLY

In Chapter 9 we looked at fabricating a cylinder assembly. This locomotive calls for those previous instructions to be repeated here, but to the dimensions shown on the drawings. There is absolutely nothing new to worry about. The lubricator sits on the end of a blind leg, tapped into the valve chest, but otherwise the territory is familiar. Ensure that the final width of the assembly fits exactly between the frames. Be fussy about this; if it is too wide, a

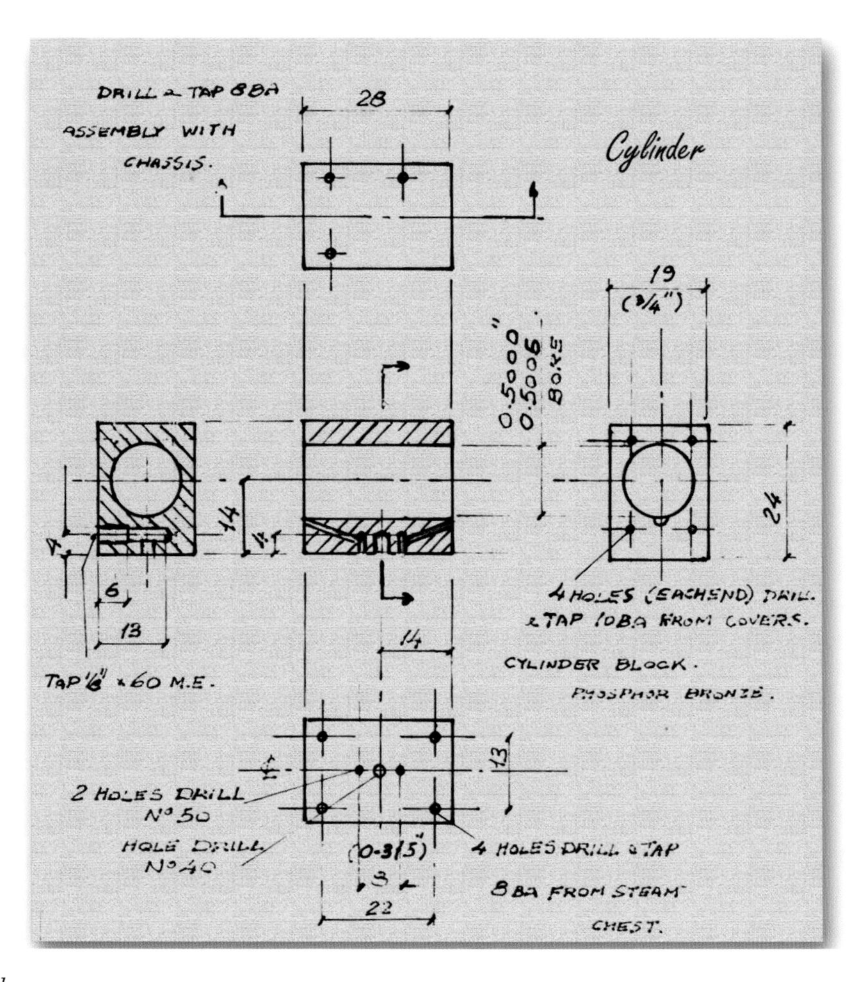

The cylinder block.

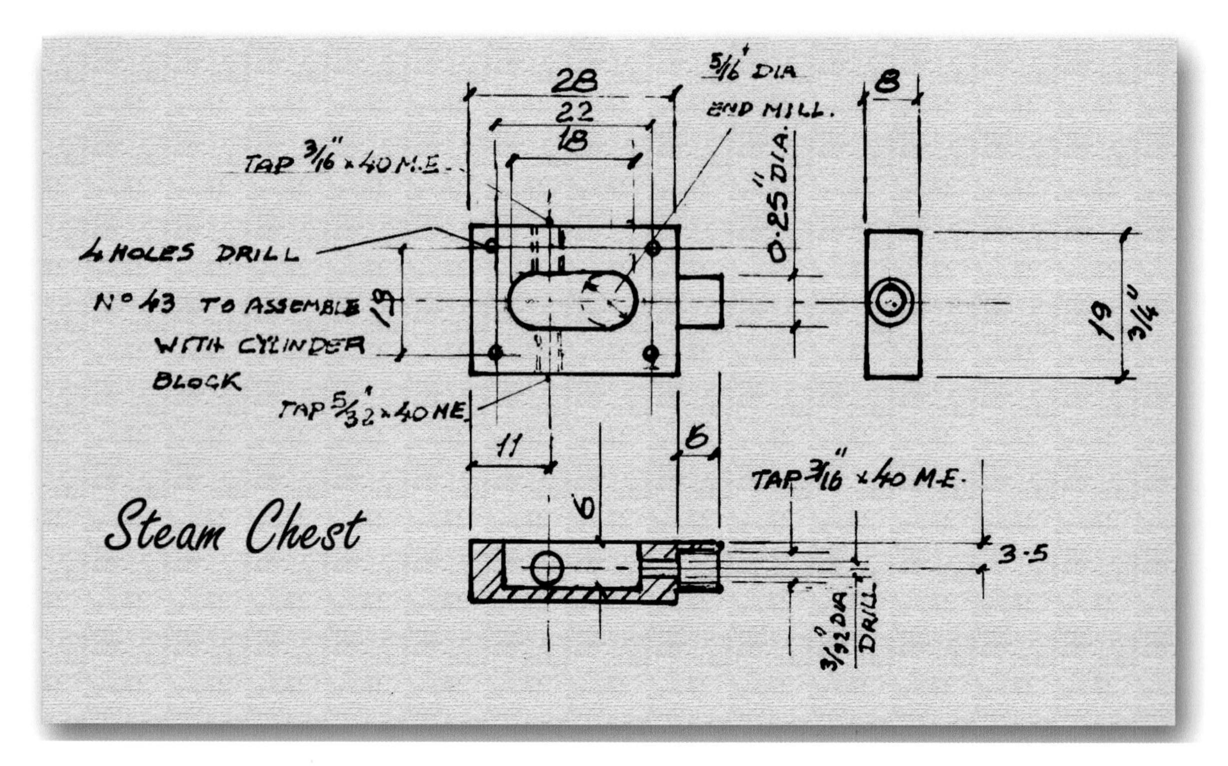

The steam chest.

small whiff can be turned off the thickness of the valve chest. If slightly too small, the difference can be made up with some gasket material as packing. However, by now the chances are that you will have got it spot on.

The crank axle, coupling rod and eccentric rod are likewise fabricated as previously described. At this stage, it is a good idea to turn up the phosphor-bronze bushes for the axles. These need to be accurately made. If you have a set of collets for your lathe then these will do the business for you. But, if not, instead of working to the very precise measurements shown, try this procedure: centre-pop some ½in phosphor-bronze rod and hold it in the three-jaw. Allow yourself about an inch projecting. Using a centre drill (Slocombe), in the tailstock drill chuck, enlarge the pop mark to a decent-sized starting hole. Then drill it through with a drill that is fractionally under ¼in. Put a ¼in reamer through. You will remember how we made a jig for turning wheels. It consisted of a cob of mild steel that had a shaft projecting, which we turned in the lathe and didn't disturb thereafter. This meant that it rotated perfectly truly. The wheel was slipped on and was held in place with a nut screwed onto the end.

Instead of the wheel, we will slip the newly turned bush blank on and secure it with a nut, so that extending pip will need to be around 1¼in long. With the bush blank retained in place, we now know that anything we turn on the outside has got to be concentric with the inside bore. This long blank is turned until it is true. The turning process will need to be quite delicate. Then turn the narrower portion, so that your digital caliper tells you that it *just* won't go into the axle holes in the frames. Next, part the bush blank into individual bushes that are to the length shown. Using this procedure, everything will be perfectly concentric and exactly the same. Put a pop mark on the jig you made, opposite the number 1 chuck jaw. This means you should be able to put it back at some time in the future and it will run true again. Remove it from the three-jaw and attach a label to it, then put it to one

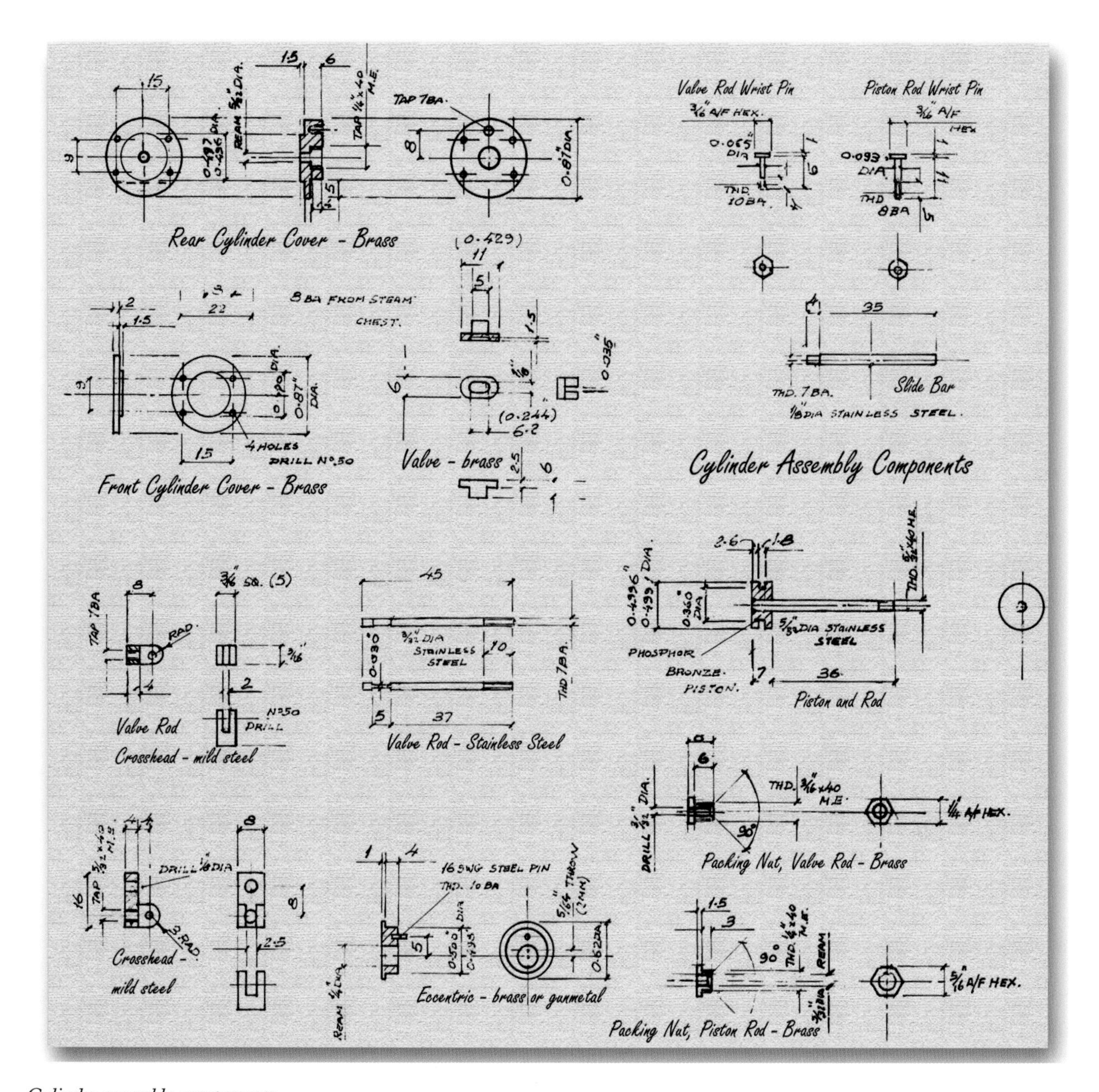

Cylinder assembly components.

side for whenever you build your next engine in this scale.

The bushes should not quite go into the axle holes. Get a taper reamer and twiddle this very lightly by hand a couple of times to put the slightest of tapers at the entrance to these holes. Take care, because a reamer can enlarge a hole at a frightening rate. When you are satisfied that the bush might nearly go through, give an initial squeeze in the vice. Note that the larger flange goes on the outside. Next, resting the job over a hole drilled in a piece of scrap sheet metal, placed on an anvil of some sort (a vice will often have one), gently tap it home with a pin hammer. No need for brute force. If, despite your best efforts, the bush is slightly loose in the frame, secure it with a dab of slow-setting epoxy resin.

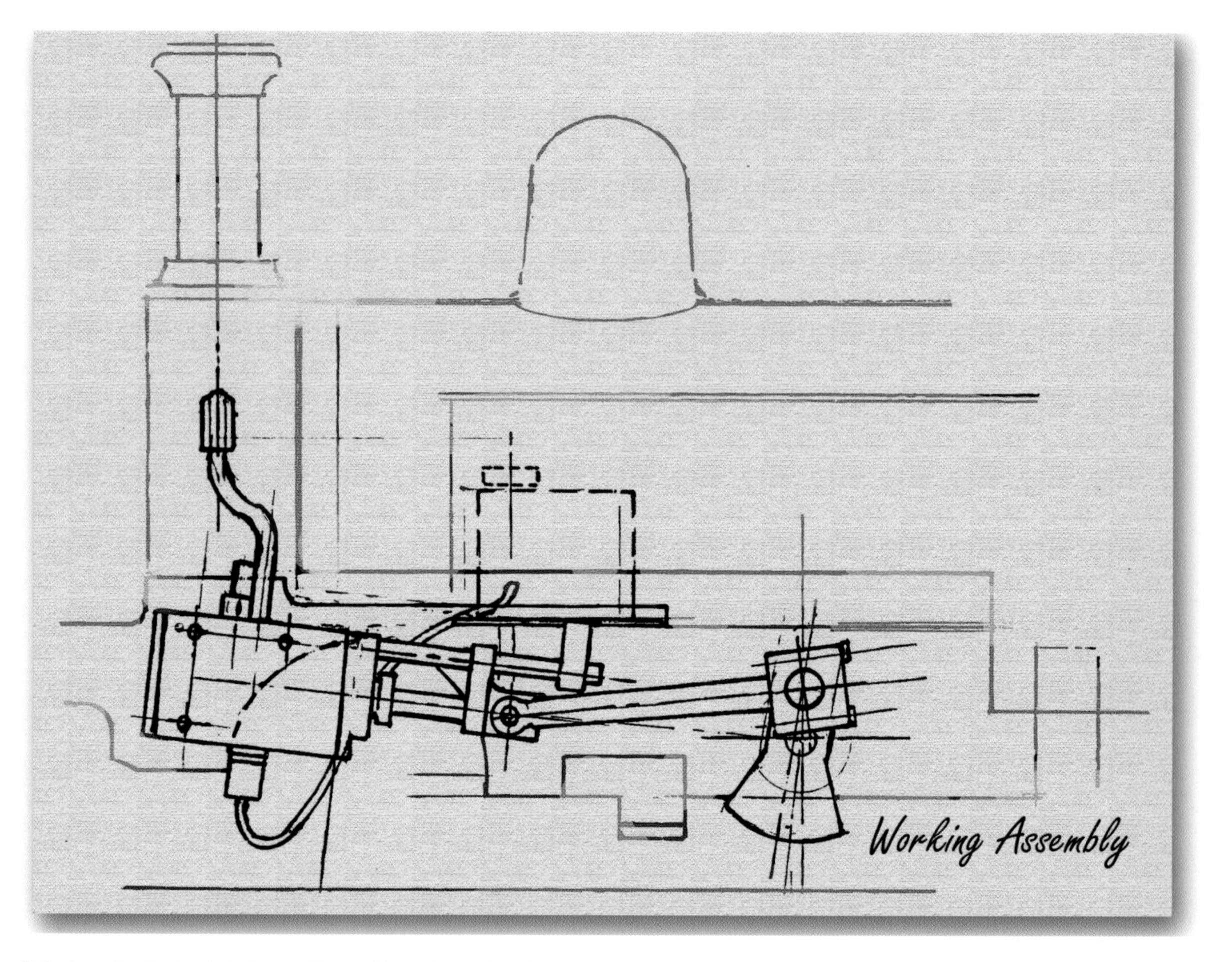

Relation of cylinder, lubricator, blast pipe and crank axle.

AXLES

The crank axle and the plain main axle should be straightforward turning jobs. The drawing calls for ¼in diameter BMS (bright mild steel). Ideally, they should be a slop-free rotating fit in the bushes. You will see that there is the faintest of tapers specified on the wheel seats at the ends of the axles. You can treat as an engineering job turning the wheel seats concentrically on axles which are already a finished diameter if you are familiar enough with your lathe. But, again, if in doubt, use the simpler but slower method of turning an axle and shoulders from one oversized steel rod. Although the outside diameter is specified as ¼in,

FITS AND CLEARANCES

Apparently vague-sounding descriptions of fits – like a tight wringing fit, for example – are in fact quite soundly based. In practice, they are much more informative to a model engineer than, say, specifying 0.001in clearance. And all the time, remember – a fit that is slightly too tight can gently be eased, but if it is loose, the part may need making again. However, because our small engines get much hotter overall than larger models in 3½in gauge and upwards we have to be slightly more generous with clearances. An exquisite model, machined to watch-making standards, would seize solid as the engine warmed up.

use the digital caliper once more to check that it is going to be the right rotating fit in those bushes.

I will draw your attention to one vital measurement for all locomotives and rolling stock. It is to get the gauge absolutely right. That may sound obvious, but it will do no harm to refer back to the first picture in Chapter 3. It shows an electronic caliper measuring, not the gauge, but the back-to-back spacing of a wheel set. It is important for this to be correct as well. Many mysterious derailments over points and crossings are due to the back-to-back measurement being slightly out.

Turn a second plain axle and we will use this as the basis for the crank axle. Cut and file the webs from 3mm BMS (this is a good example of why it pays to jump freely between sets of measurements; 3mm is a nice convenient figure). The crankpin is made from $\frac{3}{16}$in diameter silver steel. The holes for the pin and for the axle are drilled 8mm apart. The pin (leave overlong for now), axle and webs are assembled and silver soldered. You will note that the crankpin holes in the webs have slight countersinks to allow for slight 'pools' of silver solder to be formed. The excess length of the pin can be sawn off. Place the axle in the three-jaw and turn the outer faces of the webs smooth. Then saw away the axle portion between the webs. Do remember that, from now on, the axle assembly will be slightly weak across its width; as a result, don't go forcing the wheels on, at a later time, without putting some packing in the gap between the webs. This is to avoid the risk of bending the assembly inwards. File the inside faces of the web smooth, where the axle has been sawn away. Finish off with progressively finer emery paper wrapped around a flat stick.

A similar 2-4-2t in action, out of doors.

WHEELS

Now would be as good a time as any to turn the wheels, as described in Chapter 7. Note the slight taper on the tread. Make the final profiling cut on all wheels without disturbing the angle of the tool. The interface between wheel and rail is particularly important on models of standard gauge engines. It helps for smooth running. The driving wheels can be drilled for crankpins, using a simple jig. They are turned from hexagon section rod. In theory, they should be case-hardened, but in practice it would take a lot of running over the years to wear them down to a point where they needed renewing. The forces acting on small-scale steam engines are very much less than those that act on their larger, passenger hauling, cousins.

The driving wheels, when the time comes, will be a force fit on the axles, bearing in mind what we said about packing the gap in the crank axle. You will note that the smaller carrying wheels for the pony trucks are threaded $\frac{3}{16}$in × 40ME and the wheels tapped likewise so that they simply screw on. If at all possible, put these threads on using a die and tap held truly in the tailstock. If the threads are even slightly askew, the wheels will wobble. If you are at all nervous about this, consider treating them the same way as the driving wheels, with plain seats and a push taper fit.

If you are unfortunate enough to have a wheel that is true to its axle but slightly loose to rotate, drill a small hole that is 50/50 in the axle end and the wheel. Apply a drop of epoxy resin to the hole – as well as the entire seat of the wheel itself – and then force a tight-fitting stub of copper wire into the hole. When the resin has cured, carefully pare away any protruding bit of copper and any surplus resin.

ERECTING THE ROLLING CHASSIS

We are now going to dismantle the existing frames so that the crank axle and the cylinder assembly can be sandwiched between them. We know from our first test assembly that the frames are true. There is an argument for painting the chassis at this stage,

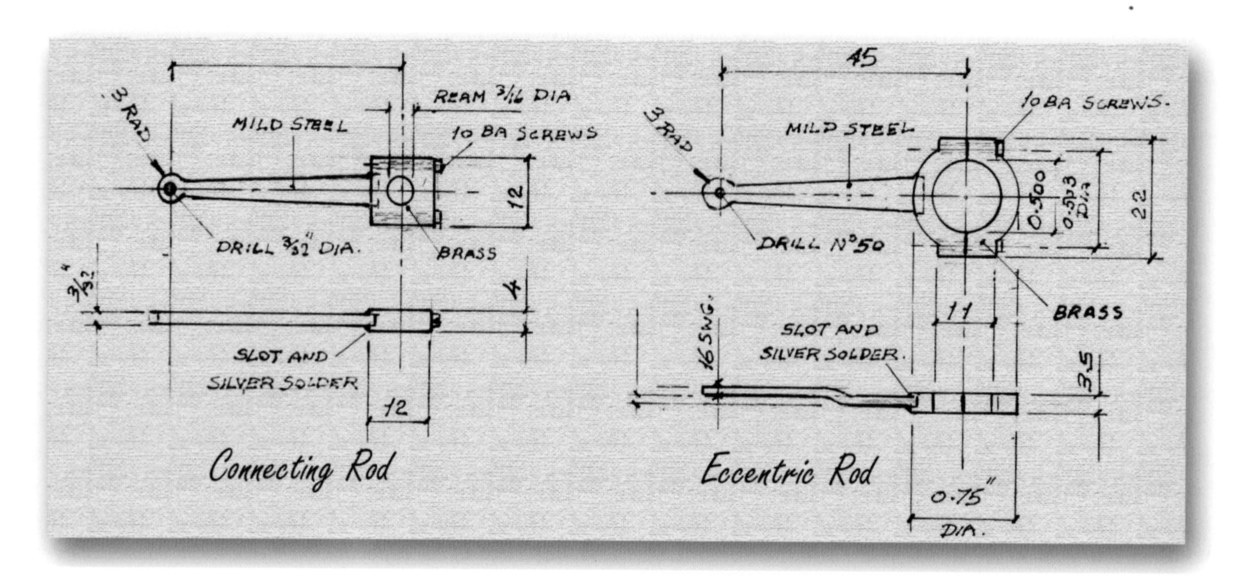

Connecting and eccentric rods.

especially if the insides of the frames are going to be red. It is possible that paint will get scratched with subsequent work, but it is still a good idea. It would be useful to clean and prime all the metal parts (not including axles and the cylinder assembly) and give them several coats of a sprayed satin black. This is easier to touch in later on. When the paint has been given some time to harden off fully, another of those milestones has been passed.

Everything is now ready to be erected. The driving wheels are pressed onto the axles, on the outsides of the frames (yes, I will repeat yet again that you should put some packing between the crank webs!). Use the bench vice as a wheel press, with some soft jaw liners in place. If a driving wheel feels stubborn, don't rely on brute force and ignorance. Carefully twist it off again and apply a further whiff of a taper reamer or even a very fine round file. It is possible to burst open a wheel casting by using the excessive force that a vice provides us with.

At the same time, remember that we have to quarter the wheels. The crankpins on one side of the engine have to be set at 90 degrees of rotation to those on the other (actually, they could both be at slightly more or less than 90 degrees, but it is essential that both axles offer the same angle). There are all sorts of jigs that can be made and have been described in print over the last one hundred years or more. I have a simple method that suits me. I make use of the two coupling rods that have been made. By ensuring that the hole centres at either end exactly match the spacing of the axle holes, we know that we can line up the wheels on one side of the chassis and connect a rod in place.

The other side will already have been set as closely by eye as possible. The second rod is offered up. There is usually one wheel that is still slightly loose on the axle, particularly if they haven't been pressed home that last little bit. By rotating a wheel on its axle, there comes a point where that second rod goes over nicely. Pushing the chassis back and fore over the bench top, it should roll without any particular sticking point (even though it is still stiff overall). If you can do this without feeling a slight hesitancy at any spot, you know things are right. You can then press the wheels fully home. You will need to insert round steel spacers between the jaws and the wheel centres (another three-handed job!), so that you don't damage the crankpins.

Knowing that the basic rolling chassis works, the bedding in of the axles in the bushes can be accelerated. This is done by that dubious practice of lubricating the bushes with a drop of metal polish and spinning the wheels against some sort of driver, as previously discussed. If you already have some Gauge 1 track to hand, just check that the chassis rolls along on it happily. In particular, check it through a point, to see that it goes across the crossing and inside the check rails smoothly. If you have followed the dimensions of the wheel profile and got that vital back-to-back measurement right, everything will be fine.

THE STEAM CHASSIS

With everything rolling smoothly, we can now button up the cylinder assembly. The connecting and eccentric rods are installed. Adjust things so that they run in and out smoothly. The little end crossheads offer a degree of distance adjustment by screwing onto the piston and valve rods. You should be unable to feel any sort of knock at the extreme ends of the travel of the piston as the crank rotates. Valve adjustment on an inside cylinder engine is trial and error, as described in the similar Wirral project. The eccentric has to be adjusted on the axle, again as previously described. When all is done, make up an adaptor for running an air line into the cylinder and carry out a brief air test. This will enable you to adjust the timing – and seeing it work will be a nice reward for your labours to date.

If you are going to run prolonged tests, make sure you regularly squirt some oil into the valve chest. Oil all moving points of contact with a light oil (sewing machine oil is perfect). A light chassis, propped up on blocks, won't run as smoothly as the finished engine on the track, as there isn't the flywheel effect of the full weight of the engine moving along to smooth things out. The cylinder assembly may still be stiff at this stage. The glands may need adjusting, but provided the wheels move evenly in both directions (after you slip the eccentric stop round by hand), you know you have a completed 'air chassis'. There is still a very long way to go, but this was the most difficult part done.

PONY TRUCKS

In this design, the pony trucks are very simple and should be no bother to cut to shape from 18swg mild steel and fold up. I would drill the axle holes after folding. It is worthwhile being fussy about getting these at a true right angle so that the wheels sit nice and squarely on the track. Put a block of wood between the folded-up portions, hold the job vertically in the bench drill and then drill and ream. It might be a good idea to spray paint the truck frames before the next step. The wheels screw onto the axles (although for the mechanically nervous we mentioned the possibility of pressing them). In this simple design, there is no sideways springing. The downward pressure of the wheels on the track is adjusted by slightly bending the plate close to the pivot point.

Just for your peace of mind, have a few further trial runs across a couple of points, to check that all is in order. Remember that, when the engine is complete, there will be a lot more weight bearing down.

THE BOILER

The boiler is made around a length of 20swg copper tube that is $1\frac{3}{4}$in diameter. There is also a 15mm diameter copper tube for the fire tube, together with some $\frac{3}{16}$in cross tubes. I make no apology for repeating that you should always use solid drawn tube for any boilermaking. This is what you will get from a model engineering supplier. Never, under any circumstances, use a tube that has a seam in it.

The boiler tube is cut to a length of 175mm. Take pains to get the ends square. If you have a rotary pipe cutter, use that. If not, resort to marking

The Lady of the Lake has a pert appearance . . .

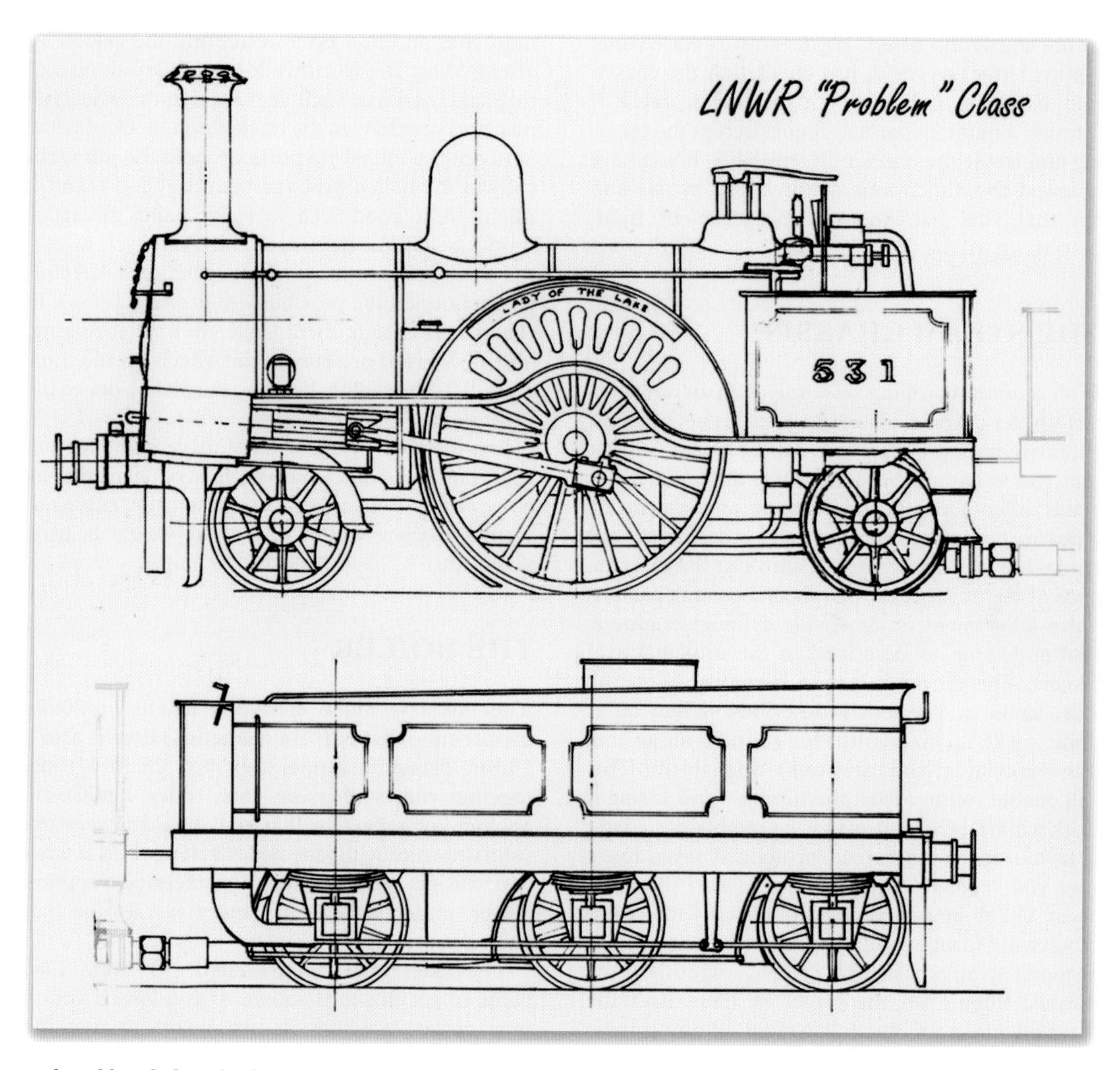

. . . but although the technology is similar to the Motor Tank, the metalworking is more challenging. Whilst the features of the Motor Tank are unfolding, here is an example of identical technology, applied to a more challenging design by Harold. You can see that the metalworking is another order of complexity. Those splashers over the driving wheels, the curved running boards and the top of the chimney are interesting. The two cylinders mean more work and there is a tender to build as well. But there is a joie de vivre *about the outline and a model of one of these looks superb in motion. In fact, there is a feeling that single drivers generally make graceful engines. Models of such things make rather fast runners when running light, but perhaps we can forgive them that.*

exactly the square ends with a pencil clamped onto a block of wood, marking onto a piece of masking tape. It is possible to turn tube off square in the lathe but it has its dangers. The tube needs to be supported by wood discs at each end, in order to prevent it collapsing and to hold it in place. Even with a tool ground to a very narrow parting-off shape, it wants to dig in. As this is such an important part of the loco, maybe it would be as well to cut by hand with great care. One method is to slip

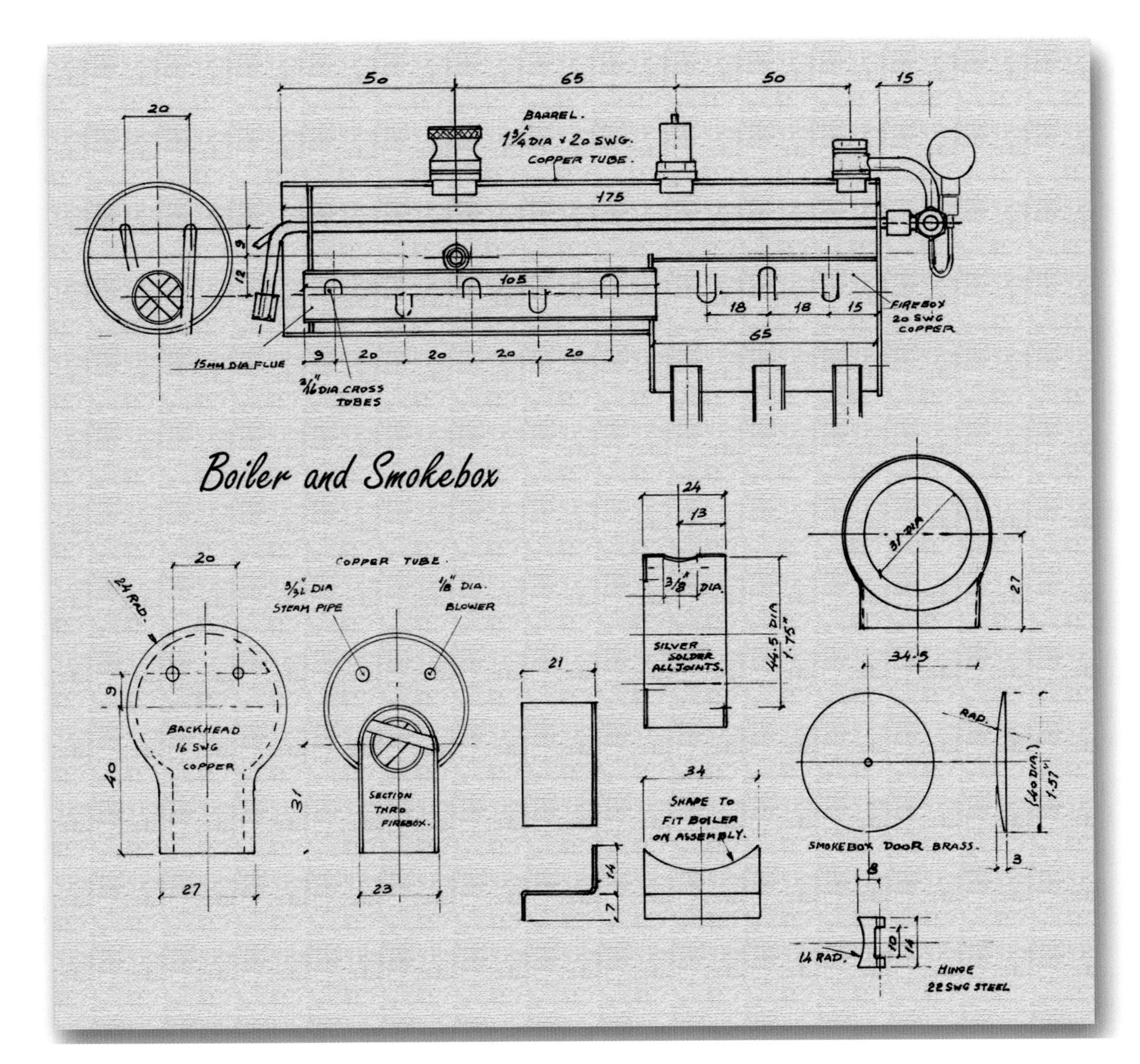

The boiler and smokebox.

a wood disc temporarily in the each end of the tube and then lightly hold it in the three-jaw, with the other end pressed against the tailstock centre, and turn the chuck by hand, whilst a pencil held in the tool holder marks it round.

Don't grip the tube in the vice, as this causes distortion. Instead, open the vice jaws enough so that the tube sits between and above them loosely. Fit a fine-toothed hacksaw blade into the frame and then saw around the mark, slightly overlong. Don't saw right through the tube, but keep rotating it by hand as you saw around the line. Use gentle strokes towards the end, without much pressure. When the end has been cut off, file the last little bit off to the marked line. A large pipe cutter can be used if you have one.

Draw a pencil line parallel to the length of the boiler, as shown in Chapter 3. Three holes for bushes have to go along this line. The dome bush hole is ½in diameter and the other two bushes are

0.38in. Grip a chunk of wooden rod – even a bit of broom handle will do – in the vice with a portion projecting sideways. Use this as a support inside the tube as you tap centre pops in. When you drill pilot holes in the bench drill, again, try to arrange things so that there is supported wood dowel going right through the boiler (perhaps propped up either side of the drilling table). This prevents the tube from buckling under the drill pressure. Open up to the correct size hole progressively. Do be aware that a drill can 'snatch' at a piece of copper, so a very delicate feel is needed as the drill goes in. My own alternative is to open out holes with a taper reamer where possible. There is a hole on the right side of the boiler for the water inlet (clack) valve.

The Backhead

This is cut from a piece of 16swg copper. A large part of it is shaped to a 24mm radius curve. To make sure you get this done neatly, you could turn up a disc of metal in the lathe and file down to it, whilst the disc and metal are clamped together in the vice. There are two holes, drilled as indicated, for the steam pipe and the blower pipe.

The Firebox

This is the inverted U-shaped item, formed in 20swg copper. Turn a wood former to 20mm diameter, to shape the copper around. The firebox is 65mm long and stands 40mm from the top of the inside to the bottom. Cut a piece of card so that it will shape round the former to those dimensions. Mark it out on your copper sheet. If at all apprehensive, still make the copper a fraction oversize –

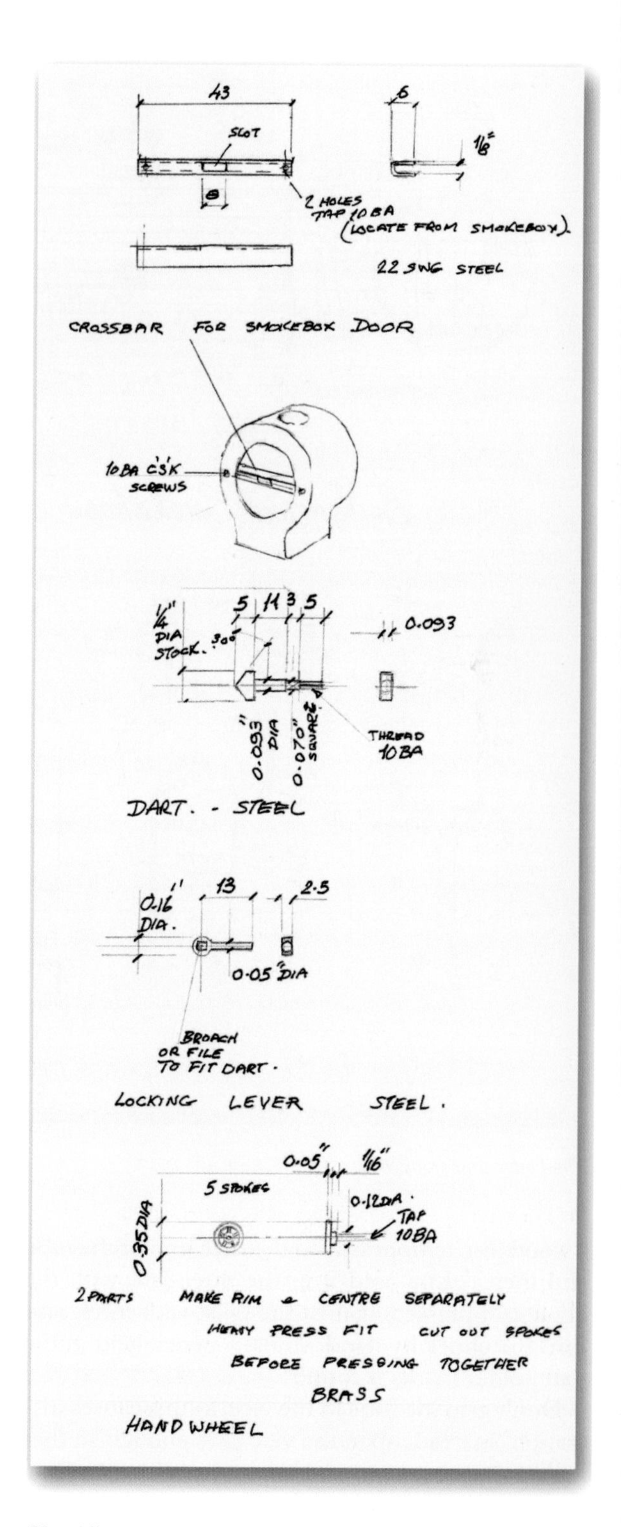

Harold's excellent sketches of the smokebox door.

SEEKING NEATNESS

Working neatly in copper is not straightforward at first. It is easy to cut things slightly crookedly (which is why I have avoided any reference to using tinsnips!) and the silver soldering can look a bit messy. Untidiness shouldn't affect how well a boiler steams, but trying to get things as neat as possible is a good ambition to have.

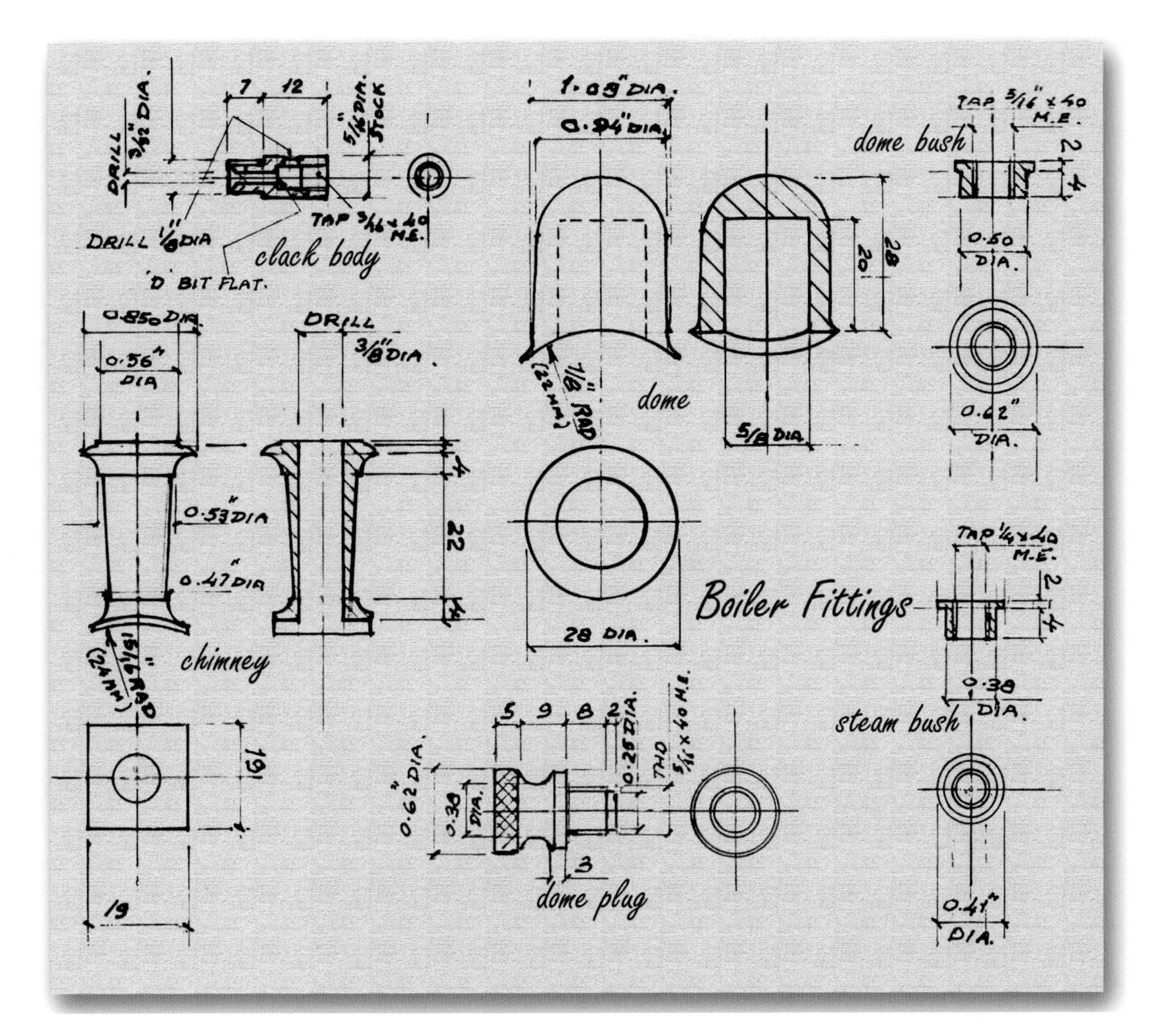

Boiler fittings.

you can always file it down when it has been rolled. Lay the copper on a firebrick in the hearth and heat it up to dull red, leaving it to cool naturally. You should be able to roll the copper with your bare hands, holding a couple of strips of wood to make sure that the rolling is even along its length. Roll it slightly too tight so that the bottom portions will spring back naturally to the vertical.

Three water tubes are shown in the drawing, running across the firebox. By now, you will be familiar with the principle of drilling holes in copper supported underneath by dowelling. There is a firebox front to make (the back is the backhead itself). This is cut slightly oversize to the profile of the firebox itself and a 15mm hole is drilled/opened out for the fire tube. The boiler barrel will need to be cut into at the rear, lower portion, so as to accommodate the firebox. It is fiddly to do with a fine metal-cutting saw blade, but avoid the temptation to reach for the tinsnips. Supporting the barrel over a piece of wooden dowel projecting sideways from the vice may help.

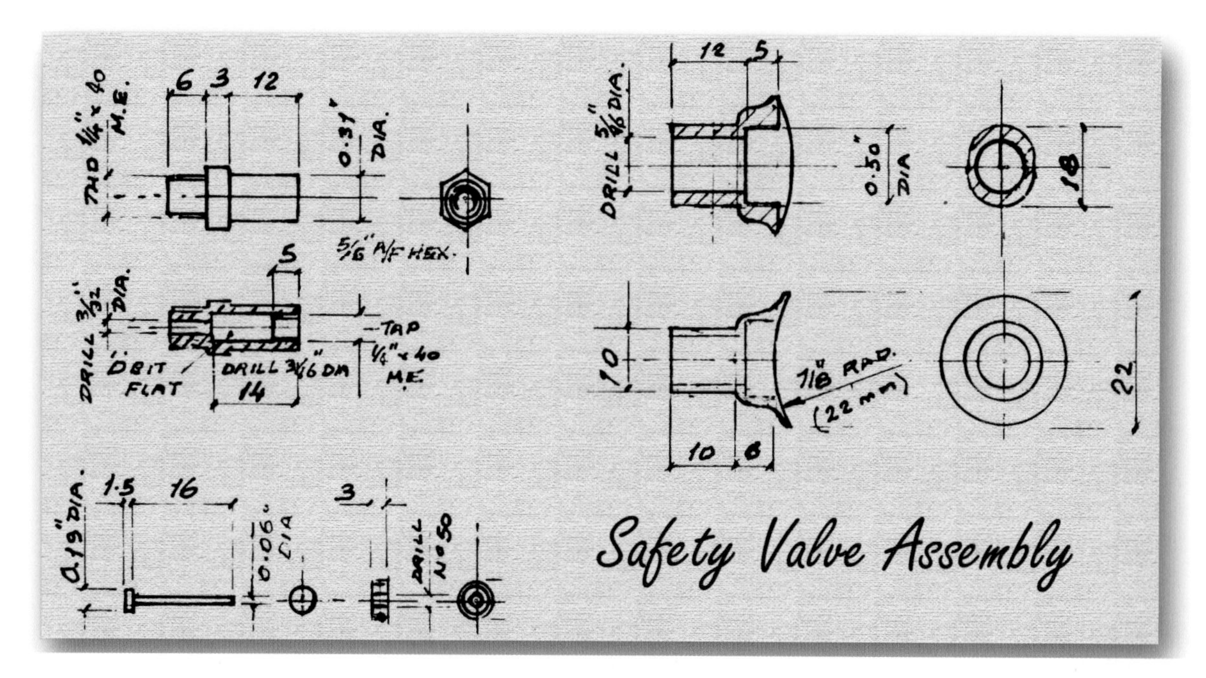

The safety valve assembly

Note that the side tanks are shorter than the length of the outer skin would lead one to believe.

The Fire Tube and the Tube Plate

The fire tube should be easier to cut square to length. A small rotary pipe cutter will do the job in a few seconds and it won't matter that the ends will be slightly dished inwards (indeed, it will be a slight help when it comes to pushing the tube through the end plates). There are five little water tubes running across. Slip the tube over a suitably sized wood dowel and drill right through by $^{3}/_{16}$in. The exact angles don't matter. The front tube plate is a tight push-fit in the boiler barrel. There is also a 15mm opening needed for the fire tube to project through.

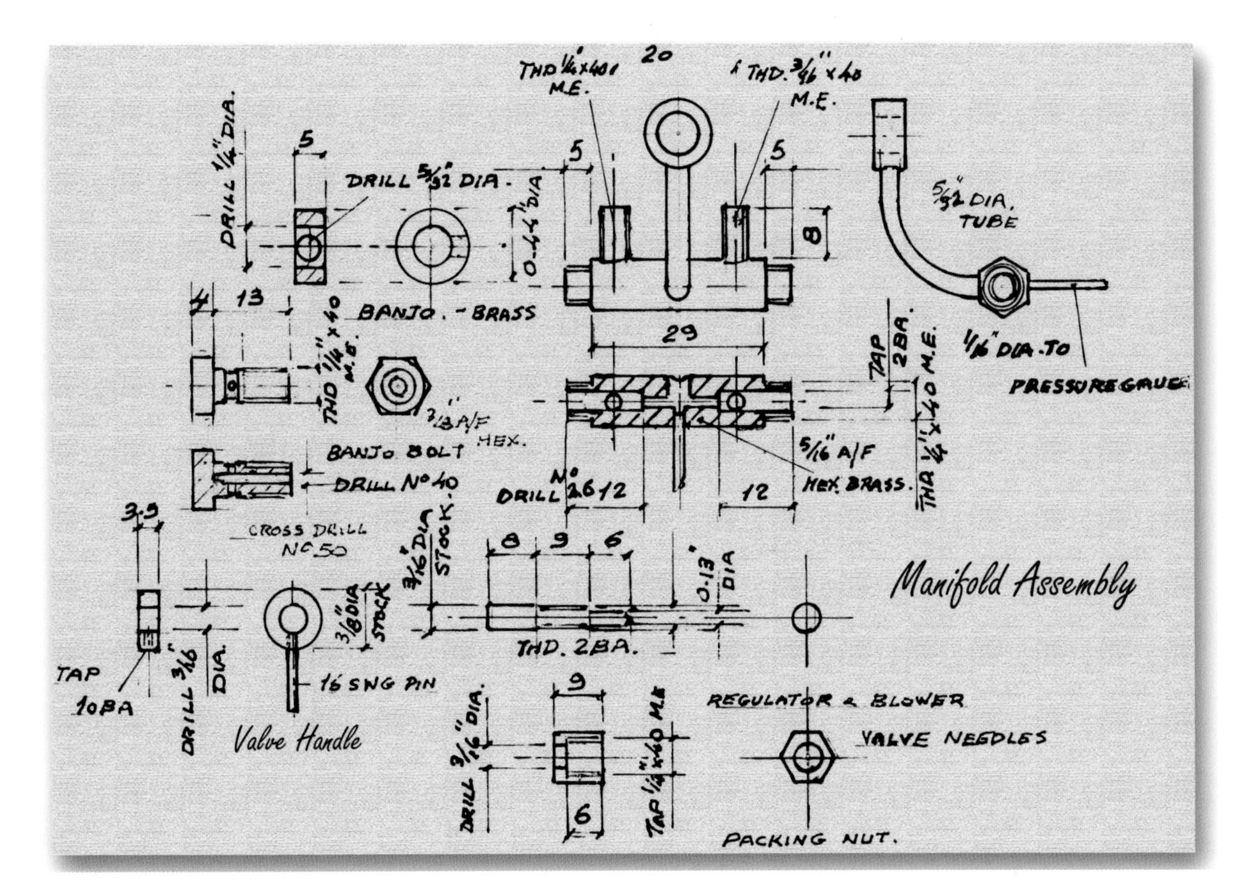

Details of the manifold assembly.

Putting the Boiler Together

Assemble all the components, including short lengths of 3⁄16in copper tube for the water tubes and the two long pipes that run the length of the boiler for the steam pipe and the blower pipe, noting that they are different sizes. Leave plenty of pipe exposed at either end for now.

Everything is going to be silver soldered, but on a job like this there is no need to use solders of different melting points. Pack areas to be joined with pieces of refractory brick. A hearth is needed and a little gas blowlamp will not be man enough for the job. You will need a moderate-size nozzle and a flexible connection to a separate gas bottle. If you can't get enough heat on the job, then acquire the wherewithal or find someone who can lend it to you.

Start by silver soldering the little cross water tubes into the firebox and fire tube. They should project slightly outside the joins and the silver solder is applied around these slight projections.

Next, silver solder the boiler barrel to the backhead. The front tube plate is soldered into place. Push the fire tube forwards from the back of the boiler barrel so that it projects slightly into that tube plate. Put the firebox in place, locating the fire tube in the hole at its front.

It is now time to join all the major components together. Try, if possible, to run the solder around everything, first time. The most likely cause of failure is lack of heat. With the job hot enough, silver solder will flow beautifully.

Push the steam and blower pipes in one end and push right through until they are ready to come out

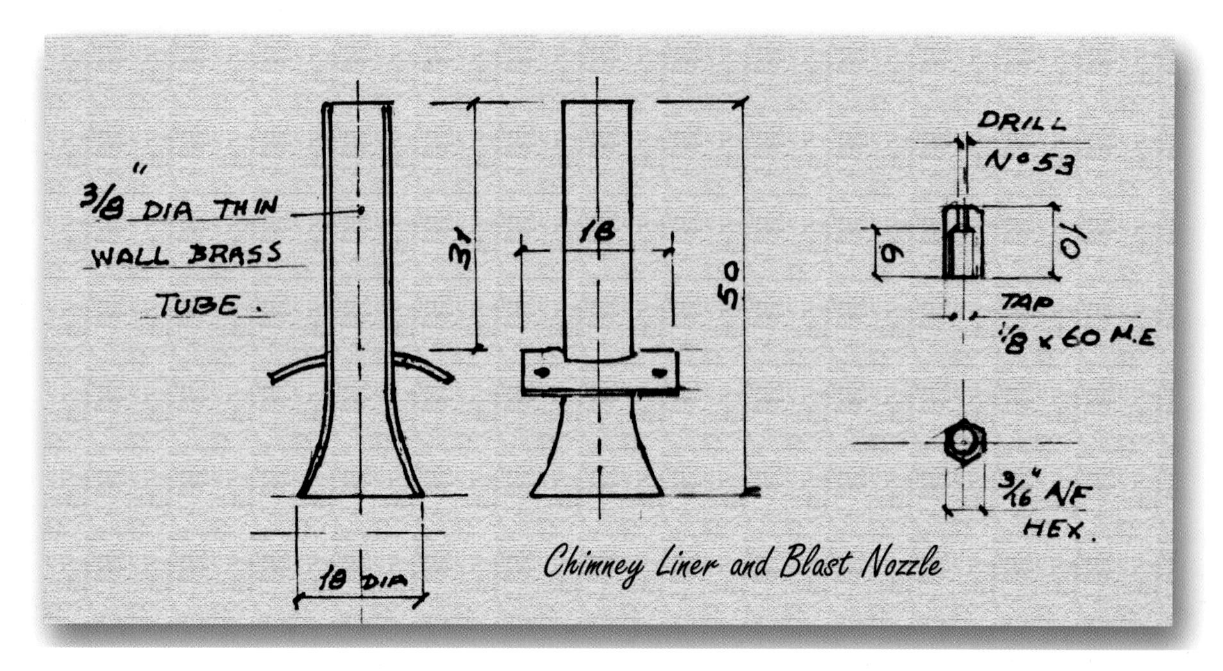

The chimney liner and blast nozzle.

> **A FALL-BACK POSITION**
>
> You may conclude that you would like to build this locomotive, but the thought of making your own boiler for the first time is a bit worrying. If this is all that stops you from building the engine, an alternative would be to prepare all the parts yourself and get someone else to do the silver soldering for you. Because all of the preparatory cutting and shaping work has been done, you could well find that even a commercial model engineer would quote a very reasonable price for the job. And you might well get the boiler tested at the same time, with a certificate issued.

at the other end, leaving plenty of pipe projecting. Things can be a bit wobbly at the far end. One dodge is to try to 'meet' the approaching tube with a cocktail stick so as to guide it through that second hole. Silver solder both ends. As well as being tubes, they act as stays, counteracting the tendency of boiler pressure to push the ends outwards.

Finally, there are four bushes to be silver soldered in place. After this, leave everything to cool down naturally and then dunk the boiler in a pickle bath for twenty-four hours. Rinse out afterwards.

BOILER FITTINGS AND TESTING

There are three boiler fittings to be made – the dome plug, the safety valve and the steam take-off fitting. There is also the clack (non-return) valve that goes on the side. These are mostly straightforward small machining jobs. Don't rely on your safety valve being accurate with whatever spring you first use. The ideal spring for the safety valve would be as follows: free-length 12mm 28swg wire, ten active coils with inside-diameter 0.070in, preferably stainless steel. You should be able to pull the pip at the top of the valve upwards with a pair of fine-nosed pliers. If you can do this, it shows that the valve does at least work. Temporarily blank off the steam take-off bush with a blind fitting and a fibre washer or O-ring.

Try the preliminary test we looked at earlier in the book. Take a car tyre valve and soft solder it into a second dome plug that has a hole drilled through it. If you plan to build more G1 engines to this formula, keep this adaptor bush, tie a label to it and store it in that by-now familiar box with your other jigs. Place the boiler in a bowl of water and put some air pressure into it, using a car foot pump that has a pressure gauge fitted. Being such a small vessel, the boiler pressure will jump up quickly. You are looking for any bubbles that are escaping from incorrect places, while checking the safety valve, from which they should be escaping. This is only a rudimentary test, but it paves the way for the next step. A little air compressor that runs off a car cigarette lighter socket is a useful tool to have.

Cradle the boiler sideways in the vice, filling it to about one-third. Next, put the dome plug in place (again with a soft washer). Take your blowlamp and *gently* apply heat into the inside of the firebox. Turn the blowlamp right down so that there is just a little roar. Be patient: it seems to take ages for anything to happen. You are looking to see the safety valve lift, but without any wisps of steam jetting out from anywhere else. If all is well, don't prolong the test.

Assuming everything is still in order, take the boiler along to an approved boiler tester. Enquire at the Gauge One Model Railway Association for a suitable person, who will not only give the boiler a hydraulic test to well above working pressure – using an accurate master gauge – but, being experienced in such things, should be able to answer any queries you may have about the safety of the boiler.

Further Fittings

Springing from the steam take-off bush will be what I always called a fountain, but which is also known as a manifold. It is a neat 'steam distribution assembly', which allows steam to be run through a regulator and off to the cylinder, as well as through a valve that controls steam to the blower. The pressure gauge is also plumbed into this. Although not shown on the drawing, you could put in an additional 'blip' valve to allow brief steam to be sent to a whistle. I'm not sure I would bother with such things; I find that the novelty soon wears off.

There is no water gauge fitted. Here is a somewhat heretical thought. The boiler is very strongly made and can withstand high temperatures. If you run a boiler like this dry, the blower stops working (and the wheels stop, thus ending the clear beats of steam from the chimney). The flames are therefore no longer drawn brightly through the tube and the engine stops of its own accord. It is not good practice to run an engine like this dry, but it should not cause any serious problem. If your loco stops running, it will either be because it has run out of meths or water. You soon acquire that feel for 'enginemanship', when the engine tells you what is happening. Quickly blow out the remaining flames and let things cool for a few minutes. However, after a few runs, you will develop a feel for how your engine is doing and when the water needs to be topped up. This slightly long-winded diversion is to demonstrate that a water gauge is not essential. On a crowded backhead, especially inside a cab, things are cluttered enough as it is! If you must have one, use the type that springs out on legs from the top and bottom of the boiler, projecting well into the cab area. These give a more accurate reading over a wide range of water levels and are easier to see. Ready machined kits are available for these.

The Smokebox

This is a drum type that sits on a saddle. It is of a diameter that slides fairly tightly over the end of the boiler and is rolled up from a piece of 22swg steel. The smokebox front is in the shape of a doughnut. Onto this an opening smokebox door is fitted with a hinge. Making this door is a tricky turning job for a beginner. I would put a short cob of brass in the three-jaw and then turn the curved profile roughly to shape with a round-nosed tool. The final shaping could be achieved by applying progressively finer emery paper, wrapped around a flat stick. When you are satisfied with the shape, part off the disc from the cob. Silver solder the distinctive hinge in place.

This door is fully working. A conventional dart engages in a slot in a crossbar. I noted that Harold's working sketches instantly make everything clear, so I offer them here instead of many words: exactly as they came off his pencil. I have to say that I think the five-spoked handle could be a fiddly thing for a beginner to make. Have a go, but if it is not successful, don't be afraid to buy something suitable ready made. I have to confess that, in the past, I have re-profiled an OO gauge flanged wheel.

The chimney liner is made from ⅜in thin-walled brass tube. You will see that there is an interesting flare in the bottom. Turn up a piece of mild steel to act as a former. Anneal the end of a piece of the tube and mount the other end onto a piece of tight fitting dowel. Put a couple of turns of masking tape around this be-dowelled end. It can then be gripped in the tailstock drill holder. The former is held in the three-jaw. Bring the tailstock up until the tube is just about touching the former and lock it in place. Then advance the tube onto the former with the tailstock advance screw. It will force the tube onto the former and start to stretch it open. You won't be able to do much before the tube starts to work-harden, but with any luck it will have given you the flare you seek in a single pass. Finally, cut off to length over the dowelled portion.

Congratulations, you have just negotiated your first piece of metal forming. It is a valuable technique that you could find much use for in projects to come.

The hole in the bottom of the smokebox for the exhaust pipe should be airtight. If the pipe is a tight fit in the hole, it should be alright, but if there is a loose fit it is worth sealing the gap. In the past, we would use fireclay for such things and you may well find an equivalent heatproof sealant in a shop that sells wood-burning stoves. Ordinary epoxy resin would work, but would make dismantling more difficult should the need arise. The nozzle for the blower and the exhaust pipe should be as closely aligned as possible to the centre line of the flared chimney liner. This maximizes the sucking effect of the hot air being drawn through from the burner.

The blast pipe from the cylinder needs to be aligned under the centre of the chimney liner. The blower pipe also emerges from the front of the boiler, but is not shown in the drawing. It is annealed to make it easy to bend upwards so that this also projects near the centre of the liner. A nozzle is formed in this by squeezing the end of the pipe round a Number 60 drill or equivalent. This alignment is fairly critical in how well the hot gases are sucked through the fire tube.

The Lubricator

This is an interesting little brass-smithing job that should present no problems by now. It is a small box with a screwed bush and filler cap on top and a ⅛in pipe emerging from the bottom. Make sure that the pipe is well annealed so that it can be gently bent between the fingers when it comes to be fitted.

Boiler Furniture

The chimney, dome and safety valve cover are fairly conventional, but are a reasonable sliding fit over inner components. We have spoken elsewhere about turning a thin flange to such things, annealing the metal and then hammering the part-machined component over a round bar. This is the easiest way for a beginner to achieve the 'saddle' effect. The non-return valve (often called a clack valve) is a typical example of a large family of such things. The principle is to have a small ball that can move in a tiny enclosed space. Boiler pressure forces the ball back against a seat and stops water or steam escaping. But a greater pressure, as from a pump, will force the ball off its seat as water in pumped in.

There is an Enots valve fitted adjacent to the right running board. This is screwed upwards into a 'post' made from ⅜in diameter brass, which forms a rigid anchor point. I could write a wordy description of how things go together, but things are much clearer with another of Harold's splendid little sketches.

If I were making this locomotive in 2.5in gauge, I would fit a conventional hand pump in one of the side tanks. The top of this would have a slot with a hinged lid. When open, an extension handle is slipped over the pump to work it to and fro. The side tank on the other side is just a plain tank. Between the two is a connecting pipe, running under the boiler, called an equilibrium pipe. A common wheeze is to have merely a couple of stubs of tube emerging from the tanks, over which slips a length of plastic tube. Both tanks would need to have removable filler caps instead of dummy ones.

THE SPIRIT TANK

The three burner tubes are made from thin brass tubing ⅜in in diameter. Do not use thick tube because it builds up heat. A spirit tube is made from ³⁄₁₆in brass tube. You will need a plug on the end, threaded 6BA. This is to screw to a support that stops the assembly from drooping downwards.

The burner tubes are drilled ³⁄₁₆in near to one end and thin brass caps are silver-soldered in place.

HEAT BAFFLES

I have a personal liking for a very easily made addition to the small spirit tube. It consists of threading alternately sized brass washers (both having a ³⁄₁₆in centre hole) onto this tube, thus forming a finned surface. The washers are tacked in place with soft solder. These fins help to disperse heat from the burners as it tries to conduct its way towards the chicken-hopper sump. They also provide drip points to stop any spilt meths from creeping towards it as well. They are purely a quirk of mine and are not specified in the drawing; the loco will manage perfectly well without them. Note that there is often limited clearance in these small scales. If the little fins catch on crossing rails on points, file them away from underneath.

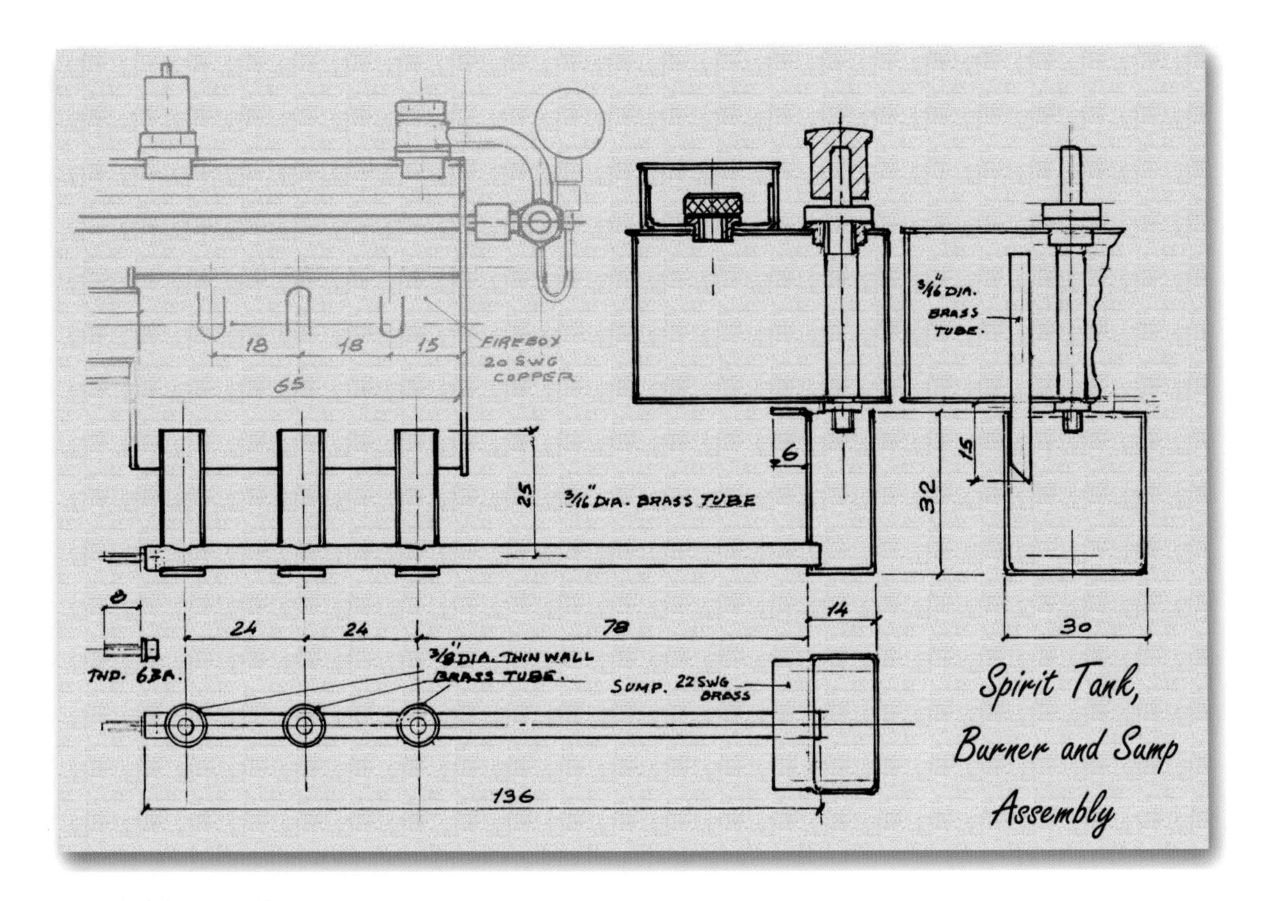

The spirit-firing assembly.

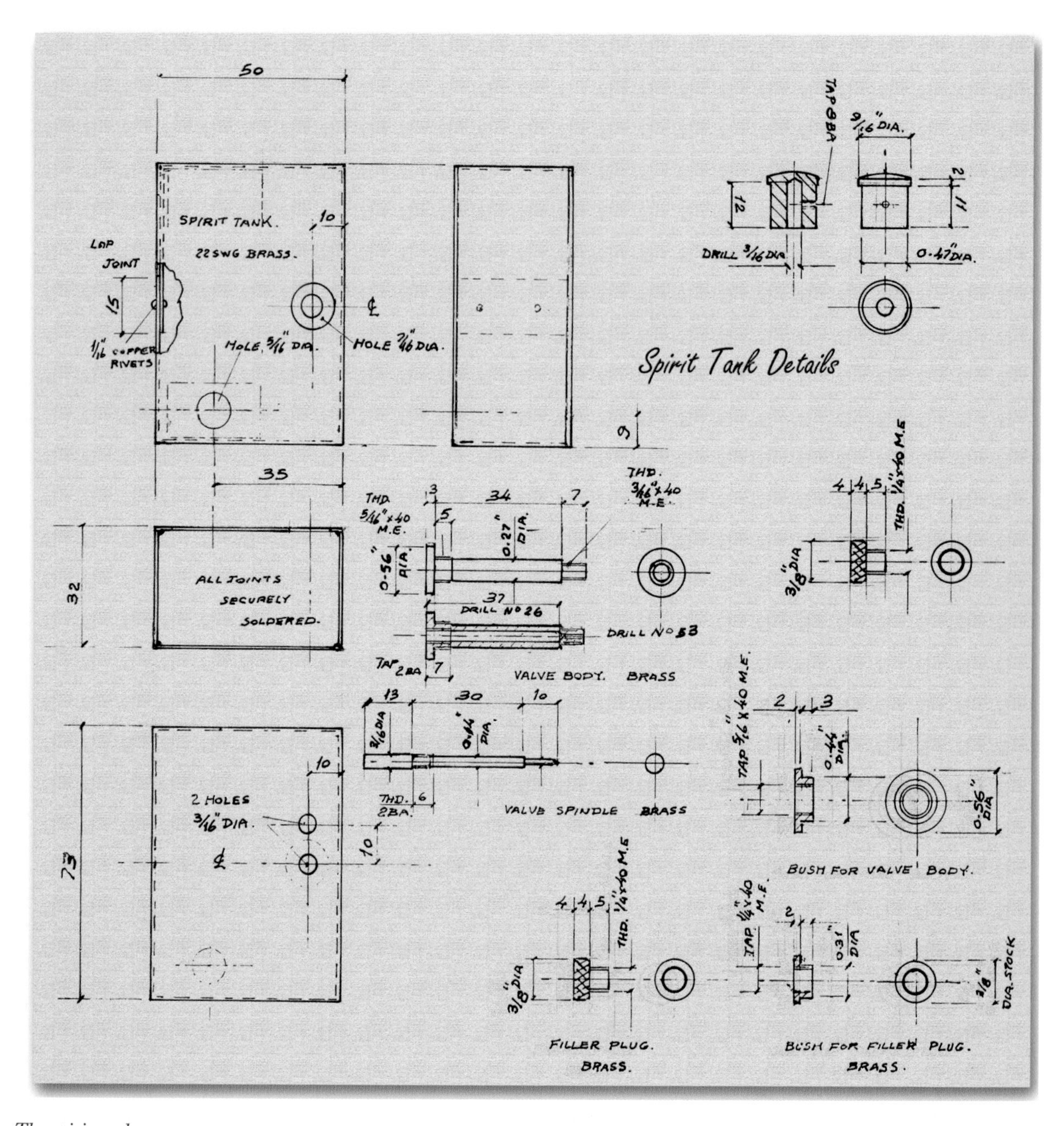

The spirit tank.

Three small holes – say, 1/16in – have to be drilled through the pipe so that each of these holes will be inside a wick tube. The wick tubes can be soft soldered onto the feed pipe.

The sump is a small open-topped box made of 22swg brass. I suggest folding up the four 'walls' of the tank from a single piece of material. One useful tip is to scribe the fold lines quite substantially, using a Stanley knife as was done for Dacre components way back in Chapter 5. These then act in the same way as etched folding lines would in a brass kit and make the folds neat. When you cut out the blank for these four 'walls' note that there is an additional projection upwards on one of them to be bent to form a locating lip. Soft solder the join of these folded four walls. I would make the 'floor' of the box slightly oversize so that it too forms a small cooling fin at the bottom. However, the real advantage is that it is easy to soft solder this to the walls.

Soft solder the spirit pipe to the three wick tubes and the sump. It would be a good idea to find a suitable slug of metal to clamp to the sump, in order to prevent that previously soldered seam from springing open when you solder the steam pipe. Because all this metal is quite thin, you can use a 50W or larger soldering iron for assembly, rather than resorting to a naked flame.

Having managed all of the foregoing jobs, you will find that making the actual spirit tank is straightforward enough. It is basically just another plain 22swg brass box. The drawing shows that a single wrapper forms the four walls. This time there is an overlapping joint to be soldered in place, reinforced with a couple of 1/8in copper rivets. The top and bottom pieces are drilled as indicated. But before assembly, there are a few simple brass items to be prepared. You will see that a 3/16in brass tube, with an angled end, projects downwards. The amount of this projection is *absolutely critical.* It determines the height of the spirit in the wick tubes. If it is too short, spirit will spill over the tops of them. Do remember that things like the filler plug need to be airtight, for the chicken-hopper effect to work, so use a fibre washer.

The fittings are simple enough to make, but you will see that the filler plug is shown as having a knurled grip. If you haven't got knurling tools, there are several alternatives for a non-slip finish. One favourite is to mount a fine pointed tool sideways in the tool post. Catch the plug in the three-jaw and then advance the tool sideways – without the lathe being switched on – just across the edge, so that it cuts a single groove. This is a very simple example of planing. Slightly rotate the chuck so that you can plane another line next to it. Then keep going round so that you have these planed lines all the way round. They won't be perfectly spaced, but you will have that non-slip finish you were looking for.

If you wanted to make them accurately spaced, remove any cover from the gear train of the lathe. Select a large wheel and then rig up some simple home-made device that allows you to drop a small wedge-shaped blade onto the gear wheel between two teeth. This means that you can rotate the chuck by hand an even distance every time. This is a simple form of indexing.

Alternatively, if you don't want any of this fuss, simply roughen the area of the plug required with some very coarse emery paper. Or, simplest of all – just leave it plain!

You will see that the bodies of the regulator and blower valves are drilled with a 53 size drill. The valve spindle is turned to just 1mm diameter at the bottom. The principle is that it seats down in the hole when screwed shut. Use a small, sharply pointed drill at the top of that opening to countersink the hole slightly. Examine the tip of the drill through a magnifying glass and try to assess its degree of 'pointiness'. Shape the tip of the valve spindle to that angle as best as you can by eye. When you screw the spindle into the valve for the first time, place a drop of metal polish at the bottom of the valve first. Screw the spindle down tight, then unscrew it a few times; this mates the surfaces nicely.

There are two little toolboxes and their retaining clips to be made from 22swg brass. You will note that one of them is removable and conceals the spirit tank filler, while the other is permanently fitted in place.

PERFORMING A STEAM TEST

This might be a good time to make sure everything is plumbed up properly and then to try a steam test up on blocks. A refinement of this is to put together a rolling road from some steel angle and home-made grooved rollers. Things may still be a little stiff, but the engine should run to and fro evenly. As there is no flywheel effect from the engine's momentum, there may be a little jerkiness. You will need a small auxiliary blower (I always think it should be called a sucker) to raise steam. There are plenty of designs for these, mostly based on a little electric motor turning a small fan. I used to recycle metal fan-blade assemblies taken from burnt-out small domestic electric drills and the like, but you can buy a miniature blower ready-made. If the air test was properly done, the timing should be okay and the thing will tick over sweetly on steam. There should be no steam escaping when the regulator and blower valve are shut. If there is, this probably means that the needle valves are still not quite seating properly. You will have filled the lubricator with proper steam oil. Keep an eye on the oil as the test unfolds. After ten minutes of steaming, the oil should mostly have turned to water.

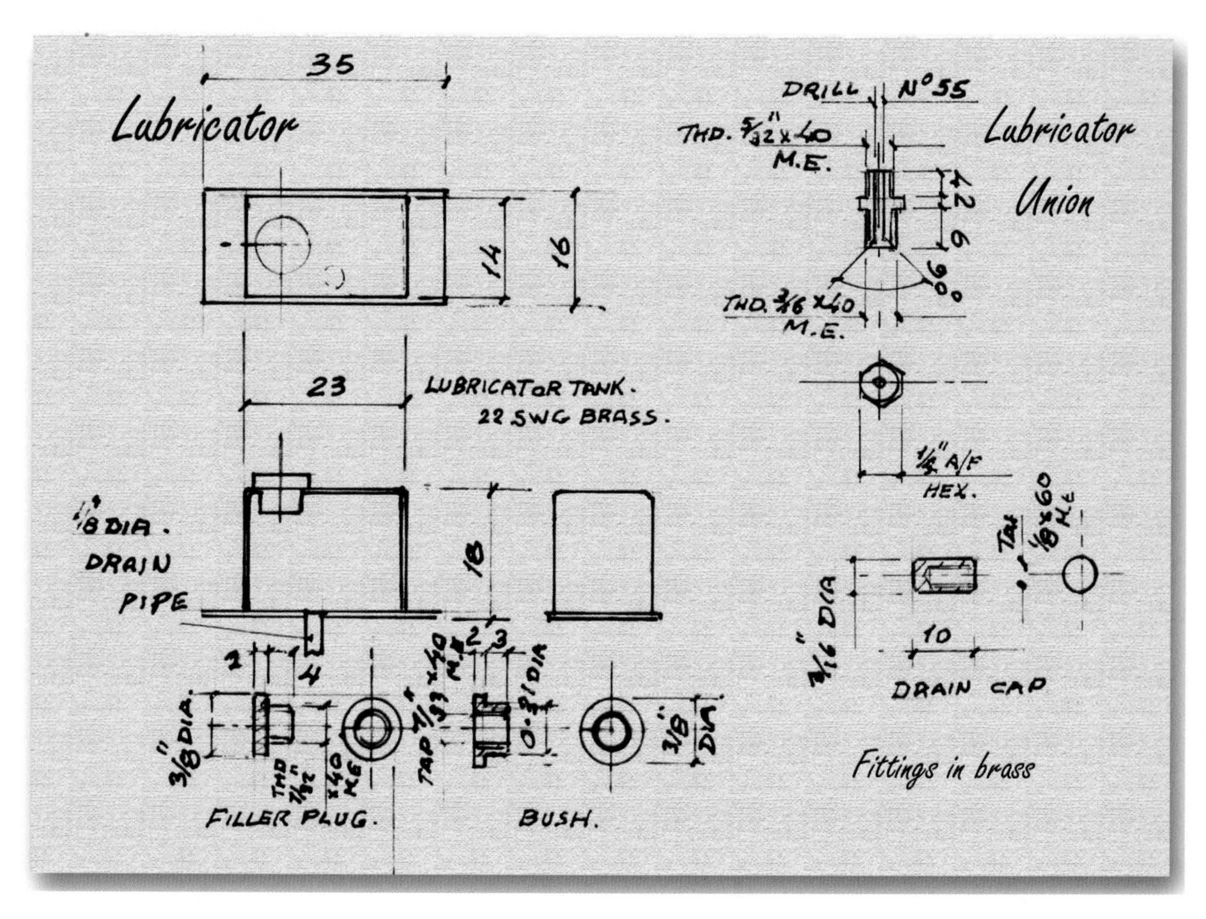

The lubricator assembly.

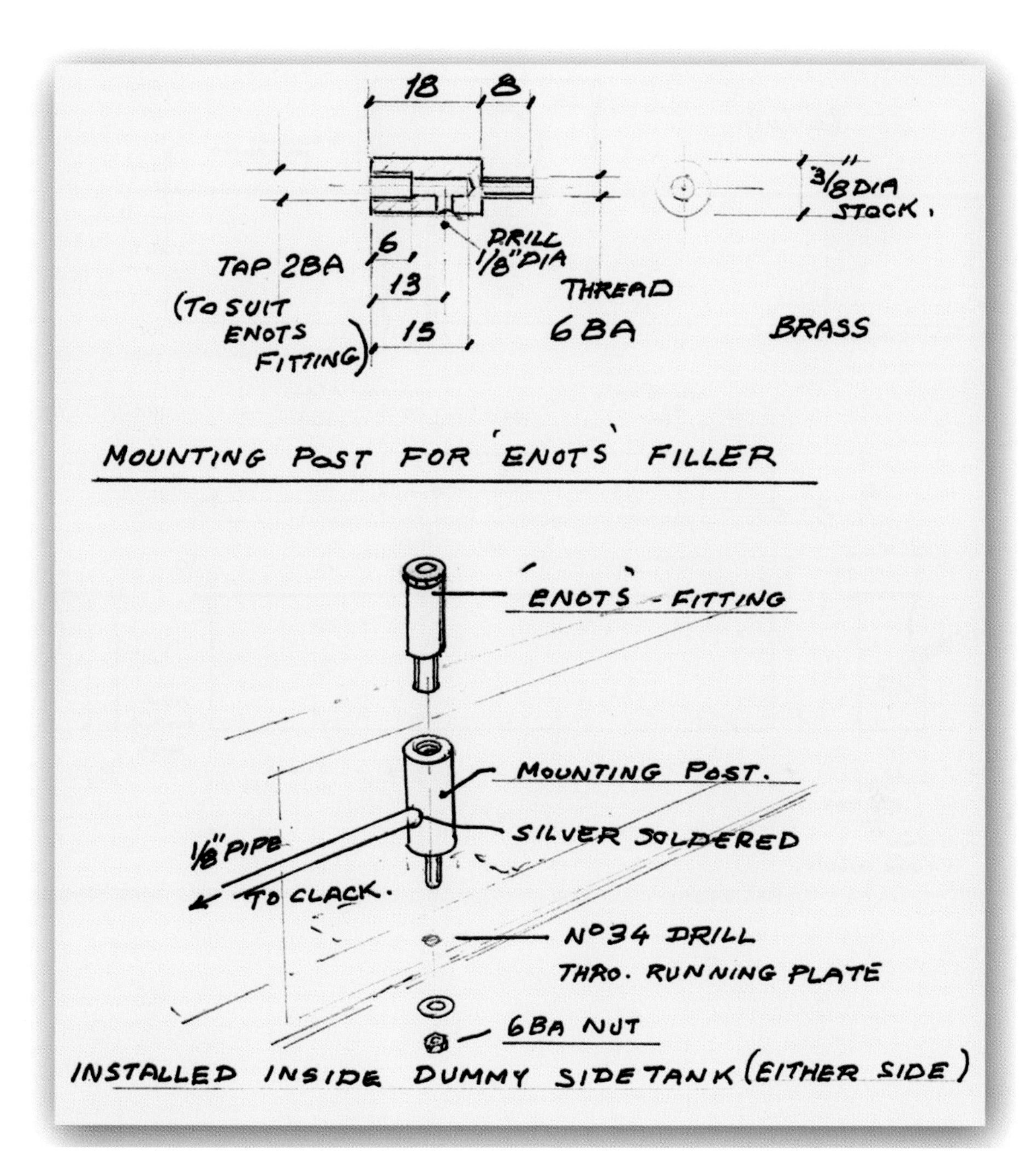

The layout of the Enots fitting – another useful sketch.

If any mechanical tweaks are needed, now is the time to do them. But if all is in order, you will be delighted at your achievements thus far . . . and rightly so. When you can tear yourself away from constantly running the engine, give it a complete clean down to remove all traces of oil, spirit and moisture, before moving on to the next stage.

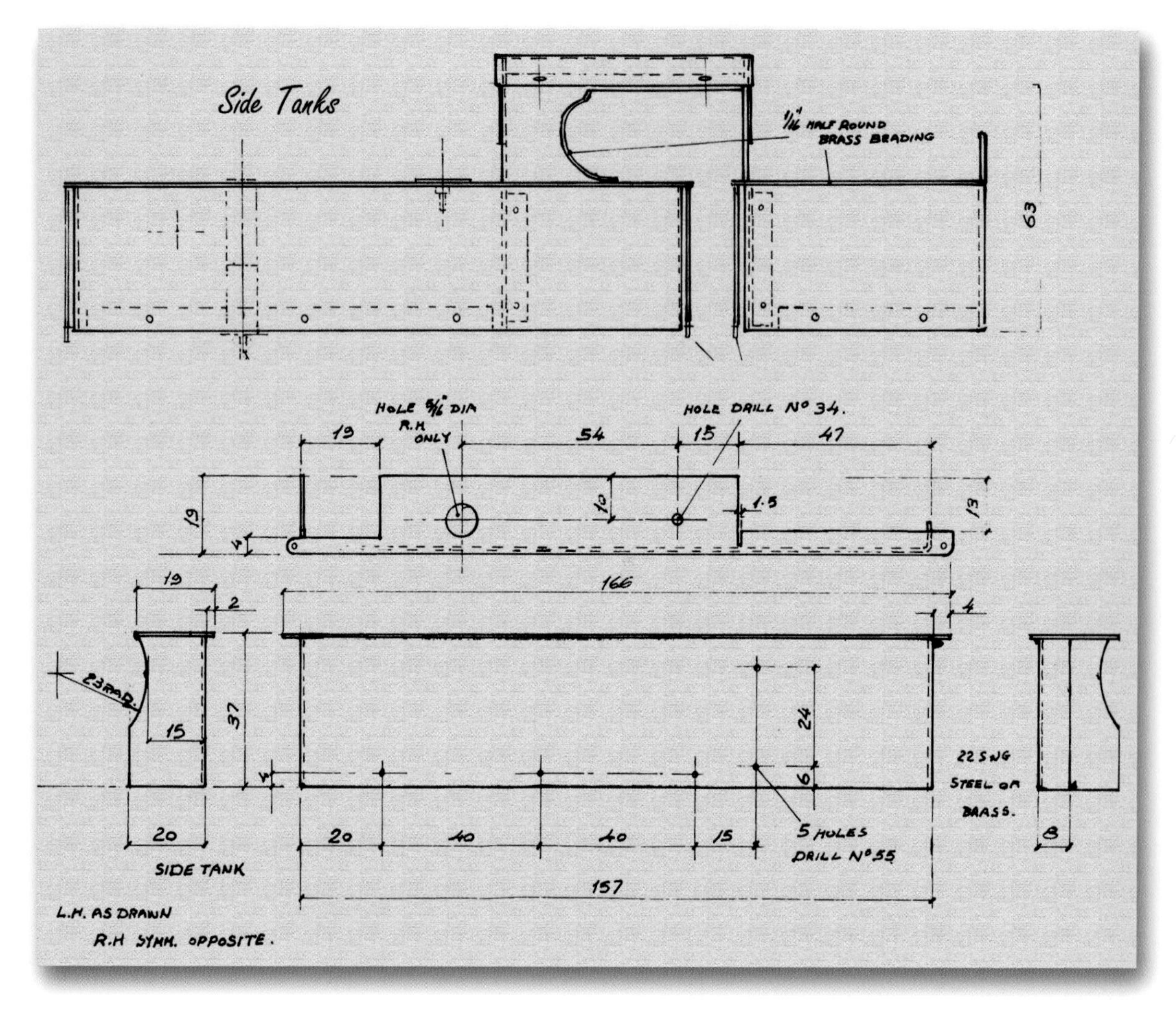
Side Tanks
1/16" HALF ROUND BRASS BEADING
63
HOLE 5/16" DIA R.H ONLY
HOLE DRILL No 34.
19
54
15
47
19
10
1.5
13
4
19
2
166
4
23 RAD.
37
15
24
6
4
20
20
40
40
15
5 HOLES DRILL No 55
157
22 SWG STEEL OR BRASS.
8
SIDE TANK
L.H. AS DRAWN
R.H SYMM. OPPOSITE.

Side tanks.

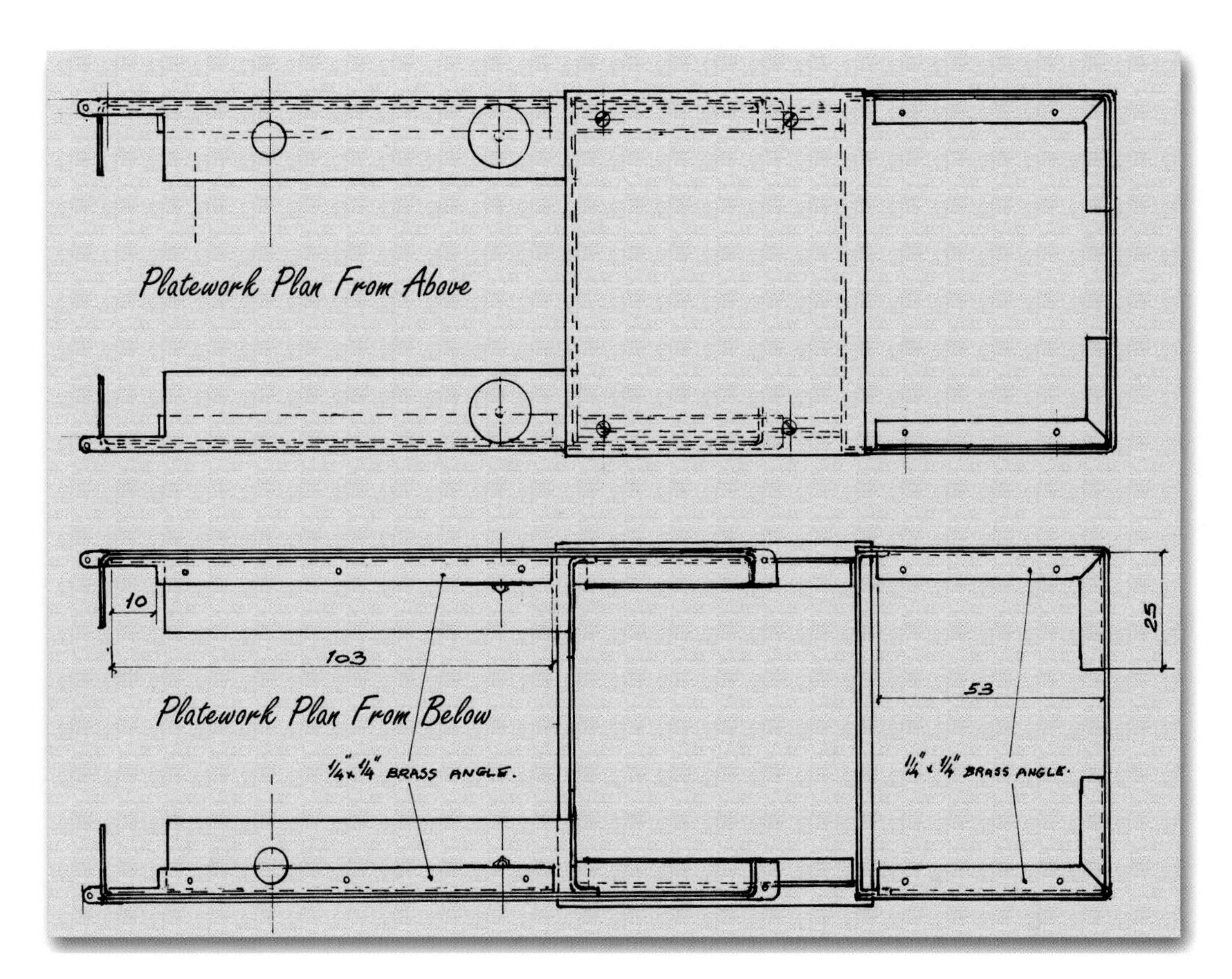

Platework plans.

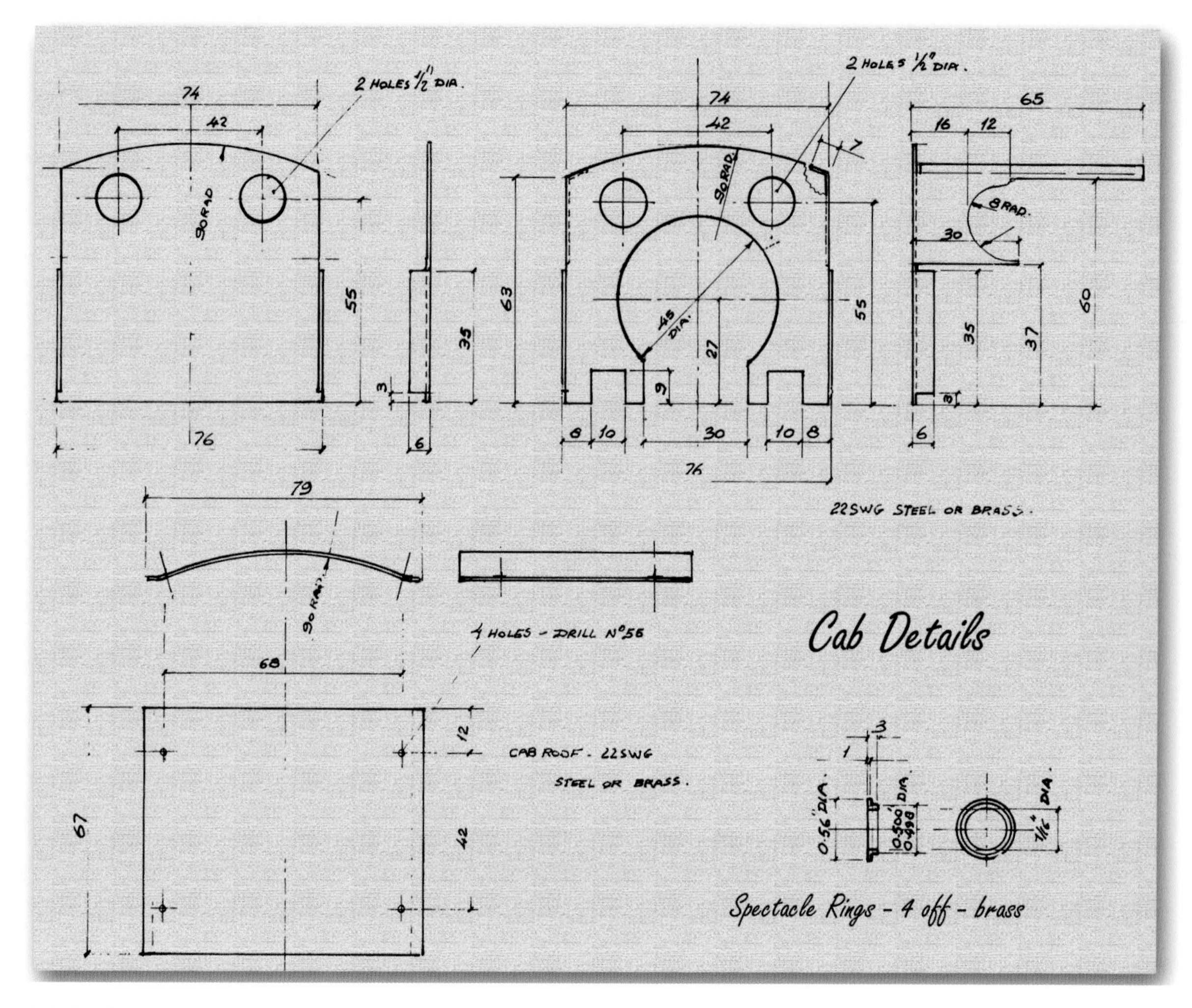
2 HOLES 1/2" DIA.
74
42
90 RAD
55
35
3
6
76
2 HOLES 1/2" DIA.
74
42
7
90 RAD
63
45 DIA.
27
9
55
8
10
30
10
8
76
65
16
12
8 RAD.
30
35
37
60
6
22SWG STEEL OR BRASS.
79
90 RAD.
4 HOLES - DRILL No 56
68
67
12
42
CAB ROOF - 22SWG
STEEL OR BRASS.
Cab Details
3
1
0.56" DIA
0.500 0.498 DIA.
7/16" DIA
Spectacle Rings - 4 off - brass

Cab details.

Case Notes:

Drummond T9

This example will be used to show how a model locomotive design, with similar characteristics to the Motor Tank, can be extrapolated from a limited amount of information. Hanging on my wall for many years was a side elevation of a Drummond T9, in its unrebuilt form. It was an old family drawing. I knew these engines well in their rebuilt form (with the extended smokebox) and always thought them the most handsome of machines. Supposing I were going to build one of these as a live steam engine in O gauge, given this side elevation as a starting point, my thoughts would run along the following lines.

Looking for a known dimension, I should find that the driving wheels were 6ft 7in diameter. At 7mm to the foot, this means that my model wheels would want to be about 45mm diameter and I would be enquiring with the Gauge O guild as to where I could find suitable castings. I would be looking for some that had the right number of spokes. I confess that if the nearest available casting didn't quite have the right number, I could live with that. This engine would be built to do a job of work in the garden, not be entered for loco-building competitions. But hopefully the exact castings could be found. The same applies to the bogie and tender wheels.

I could calculate the overall length of the locomotive from the wheel diameter. These days I would do it with Photoshop in the computer, but in years past I used a slide rule – or even sums on the backs of old envelopes. Actually, one very useful tool to have in a workshop is a small blackboard screwed to a wall, plus some chalk. I could work out the scale of my original drawing (it happened to be the right scale for a 2.5in gauge loco) and how much I would need to reduce it to bring it down to 7mm scale. If I didn't have a front elevation I could work out something from Gauge O guild dimensions. One of these engines is preserved so I could pay it a visit to run the tape over it. I would also make enquiries about published drawings. A Hornby 00 gauge model would provide a basis for the dimensions required.

This particular example of a T9 combines all of the features I liked the most: the unextended smokebox with a wingplate; the double splashers; and the eight-wheeled watercart tender. I could obtain much information about the rebuilt T9 and modify it back to the original condition. The next thing to be considered is what to do to produce a practical working model.

My own preference, given a choice, is still for spirit firing. To make it an all-weather proposition, I would choose a Smithies boiler – a boiler within a boiler. Burners would be located in the underside of the firebox area and the hot gases would be drawn forwards, past the inner boiler and out up the chimney. As an alternative to chicken-hopper spirit tanks, an unorthodox approach would be to make a very large but thin tank under the tender floor, between the frames. This is such a big tender that a generous amount of spirit could be carried this way. One radical ploy would be to make the space between the tender frames a complete spirit tank, with the axles running through 'tunnels'. The body of the tender could contain water and a hand pump. I make no claims that this is the best way to do things; merely my option.

There would be a single chunky cylinder between the frames to drive the front driving axle. This is simple and rugged. If the loco was going to tackle really sharp curves, I would narrow the mainframes in the area above the front bogie, which would keep the profile of the frames looking correct. The usual alternative to

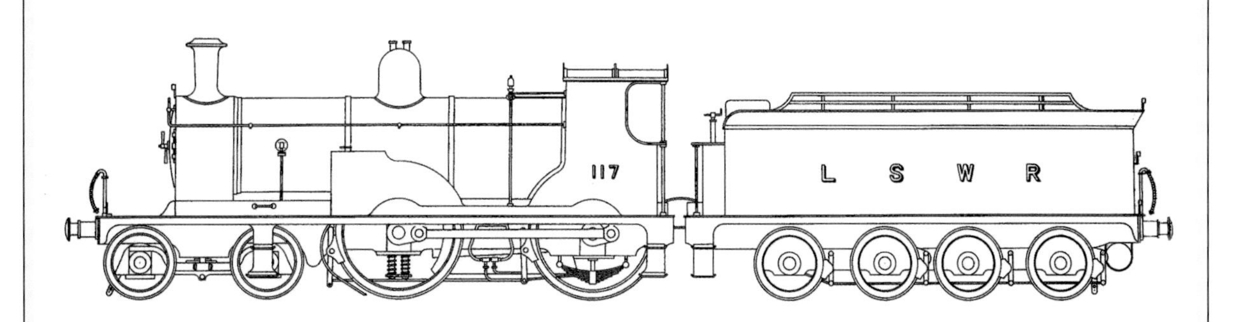

The Drummond T9 in original form. Drawings are available and one example (albeit rebuilt) has been preserved.

Drummond T9 *continued*

this is to dispense with the frames altogether in this area and let the bogie swing freely. But I would prefer to avoid that fresh air look if at all possible. Those steps adjacent to the rear bogie wheels might cause trouble on very sharp curves and, in such circumstances, I would reluctantly leave them off. I would be inclined to make the tender frames slightly narrower than they should be, to allow a little sideplay in the tender wheels. The large tender would also offer plenty of room to house a radio receiver and servo to operate a steam valve. If the need were for a rugged and reliable engine for use out of doors, I would leave off the brake gear and sandpipes.

In 2.5in gauge, there would be room and opportunity to make a more 'serious' model. There would be two cylinders between the frames, so even if slip eccentric gear was still employed, the engine would be self-starting. Simplified Walschaerts gear would allow for reversing with radio control. There would be room for gas firing whilst still having plenty of room for water in the boiler. Indeed, this engine could be coal-fired and haul full-size passengers, but by now we have begun to cross the line away from the simpler world of locomotives intended for scenic garden railways.

The big tender would also offer plenty of space for radio control. These days, it has become a cheap add-on option. My only personal addition to the large pool of information that is freely available is to suggest that a live steam engine should not need radio contol to enable it to run smoothly. Aim to build a locomotive that behaves itself manually first and foremost.

The elegance of a T9 works well in any scale.

THE PLATEWORK AND FINISHING TOUCHES

One of the advantages of this prototype is that the platework is extremely simple, consisting of rectangular shapes, mostly bolted (10BA countersunk brass) together using brass angle. Again, I would pilot-drill the cab windows and then open them out with a taper reamer to prevent the risk of the job snatching; I would do this before cutting the sheet to overall shape. After having tackled all of the stages in building this loco so far, these last jobs should present no real problems. You could take the easy way out and buy ready-made spectacle rings (also known as cab lookout bezels), or there might be a model ship's porthole ring that would fit the bill. But if you particularly want to make your own, drill the opening first in an offcut of oversized brass rod. Force a piece of wood dowelling through it, chuck it accurately in the four-jaw and then turn outer profiles for all four. Part them off whilst the dowel is still there. Then you just have a small sliver of wood to press out from each ring.

The buffers are simple enough, as are the dummy tank fillers. The dummy spring assemblies are fabricated, mostly from strip. They are fiddly to make but do not call for anything that is really difficult. When soft soldering the parts together, try to keep the solder on the back face of the spring where any untidiness will be hidden by the smokebox.

There are four dummy brake hangers shown. There would appear to be no easy dodges here to assist in making them as per the drawing. They have to be filed up neatly from 16swg mild steel. I suggest making a template in thin card first, with the hole drilled through. You could drill a hole somewhere in your little piece of sheet steel, then saw and file around it. If you have patiently shaped the item first and then fail to get that hole drilled exactly in the right spot, it is disconcerting to have to scrap it and make a replacement. Earlier in the book we looked at making brake shoes out of large thin hex nuts. You might conclude that it would be just as easy to turn the brake hangers without the profile of the shoe, then add a proper shoe afterwards. On the other hand, you might decide to omit these brake assemblies altogether for a first engine.

Making the coupling hooks is a straightforward job, or you can buy ready-made ones over the counter. This gives the option of having proper screw link couplings.

You will also see that two dummy brake pipes are shown on the buffer beams. I have a simple method of making these. It consists of bending a piece of annealed brass rod to a shape vaguely like a shepherd's crook. The little bent end is screwed. A couple of tiny collars are needed. I often use suitably sized olives from steam unions. One collar is slipped on and secured near the top with a dab of soft solder. Then a suitably sized tension spring is slipped on (I am a compulsive hoarder of springs recovered from old video recorders, biros and the like). The bottom collar is soft soldered on and then the excess rod sawn off. This works well in any scale. It is just a case of adjusting dimensions to suit.

There is an amount of brass beading to be applied to some of the platework with a chunky soldering iron. The boiler shell calls for some boiler

Making dummy brake pipes.

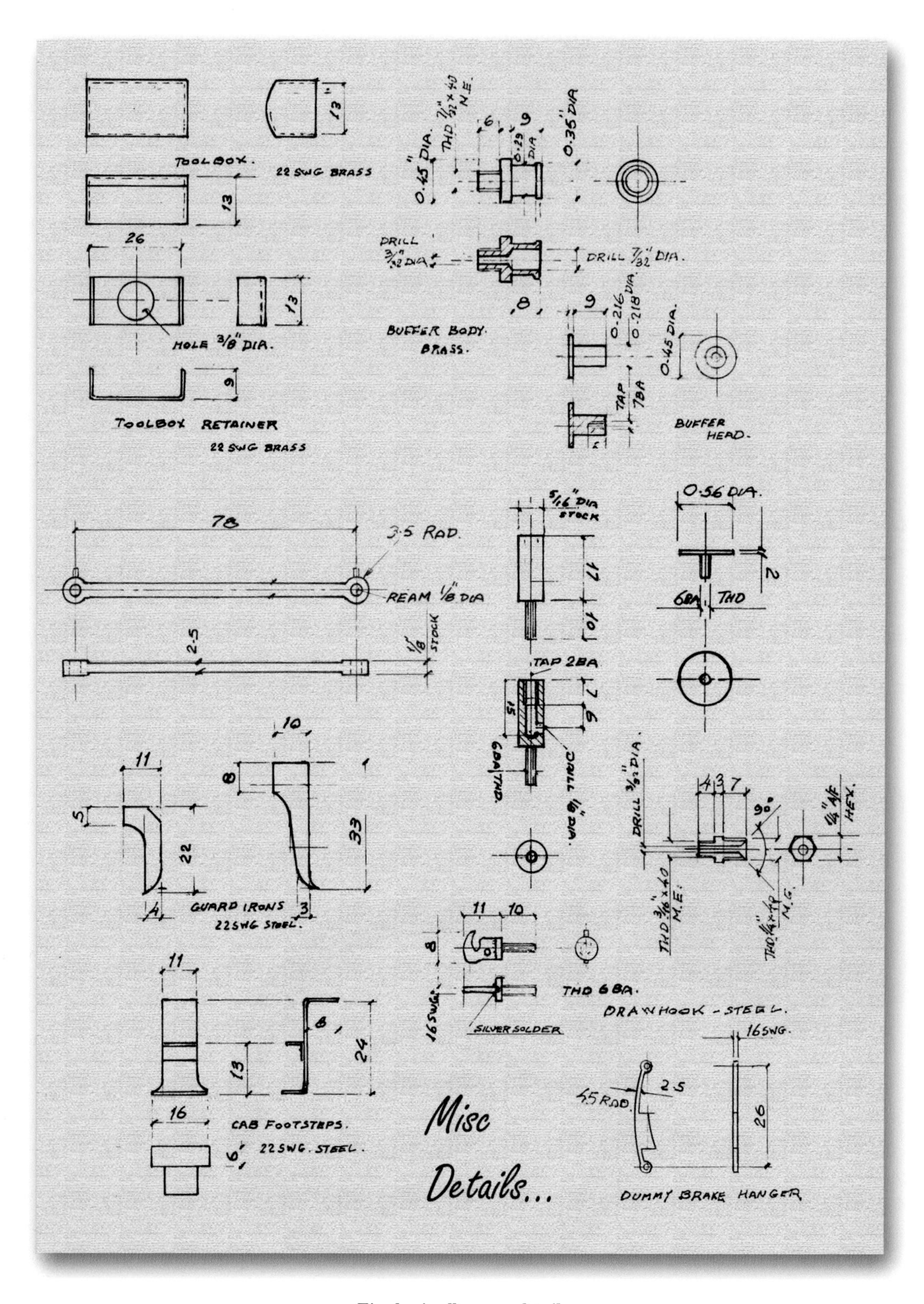

Final miscellaneous details.

bands to be added (described in Chapter 9). I will suggest that the cab roof is a 'drop-on' item. Simply shape it as shown on the diagram and then solder little brass pins to the roof, so that they will engage in holes in the supporting angle. And here is a little ploy we can use. Ideally, the curves of the roof want to be perfect to the drawing. But if the main curve is marginally too flat, the roof will need to be squeezed very slightly between the fingers to line the pins with the holes in the angle below. This makes a crude but effective 'spring locking' to the roof. An alternative is to consider making the complete upper cab as a single unit that can be lifted off entirely for access to the backhead, as per Wirral.

While the engine building was making progress, it would have been a good idea to order any etched number plates you may want. Thus they are now at hand to be soft soldered to the platework where appropriate. When the time comes for the platework to be sprayed black, let the paint go over the name plates too. They can be overpainted red by hand and, when dry, gently rubbed over with

A selection of further examples of Harold's work. And yes; the tartan livery is correct – although somewhat challenging to apply!

PACKING WICK TUBES

Wick tubes need packing with wick material. For generations this was asbestos string, but these days we have to use a substitute. Suppliers can provide a silky ceramic-based material that works well, although wick material for oil lamps can be unpicked in an emergency. Each tube is packed with strands until it is fairly tight but not jammed solid. They project upwards ¼in, looking like they need a neat haircut. Give them one until about half an inch is projecting. Splay these out slightly. If you have a tube that occasionally blows out, leave a couple of strands of a neighbouring tube a bit longer and bend them over so that they will reignite the offender. The usual cause of a failing tube is that the packing is too tight or that the hole in the feed tube is blocked.

very fine emery paper, or even an old-fashioned typewriter rubber.

You will doubtless want to run the complete but unpainted engine on a track at this stage. Enjoy its first run . . . and second . . . and third. But then give it a thorough clean, before moving to the paint shop. This has a chapter all of its own. All being well, you will have just achieved a major ambition. Internally fired engines need a blower fan temporarily placed over the chimney to get steam up. A novel alternative is to run a pipe from the front buffer beam, upwards, to form a blower. One of those cheap 12V compressors blows air into this pipe.

By following these traditional drawings, you will have learnt how to interpret them in order to produce a working steam engine. You will have become familiar with different types of dimensions and measurements. Because of this, your first engine took some time to build, but you are now equipped to build anything from any sort of a drawing, using this same methodology. Your next locomotives will be easier to build because these processes will no longer be new.

FURTHER READING

As well as studying other designs by Harold Denyer (HB Models), I recommend joining the Gauge One Model Railway Association. They have several very good projects described in detail in simple book form, such as 'The Project', an LMS 0-6-0 tender engine. Also notable is a 4-4-0 tender engine of very attractive appearance. Should you want to pursue building narrow gauge engines in 16mm, in this fully model-engineered way, then I can do no better than refer you to the works of Keith Bucklitch. He offers some excellent drawings and notes on his designs (*see* Appendices for details).

CHAPTER 11

Painting and Lining

Small-scale live steam engines present their own challenges when it comes to painting them. Because they *are* so small, they can get very hot. Every part of the model is only a few inches away from a burning flame. Combine that with exposure to oil and water, running through all weathers and a lot of handling, then it is obvious that any paint finish should be tough and durable. An ordinary painted surface will not be as durable as a commercially baked-on finish, but it will take care of most of our needs.

PREPARATION AND SPRAYING

There are steps we can take that will contribute to a good finish. The first of these is to strip the newly completed model down to as many parts as is practicable. There is a school of thought that suggests that model engines should be made as subassemblies that are fitted together with a few screws. Joints that have been soldered will probably have traces of flux left on them, which will need to be cleaned off. The usual good advice is to scrub the joints in warm soapy water using an old toothbrush, although I go a stage further. I leave parts to soak overnight in a bowl of water, with a couple of the previously mentioned denture cleaning tablets. This also helps to clean off any residual grease, including fingerprints. Thereafter I try to handle the job only by the edges or by the parts that will remain bright metal. You can also wipe the job off with methylated spirits (not white spirit, as this would leave a residue).

Steel needs to be sprayed with either red oxide or grey metal primer. Fortunately, aerosols sold for cars are just perfect for the job. The secret with all spray jobs is to keep the job and the surrounding atmosphere dry and preferably warm. I like to use a big cardboard box on its side, with a drop-down flap of plastic sheet over the open front. I briefly run a vacuum cleaner nozzle inside to suck out dust particles. Next comes a hair dryer, pushed up inside the flap, which warms the air and the parts inside. I might even have left the parts sitting on a radiator for half an hour beforehand.

The aerosol can is shaken vigorously for a couple of minutes. If the weather is cold, I put the can in

(Photo: courtesy Geoff Munday)

some warm (*not* hot!) water for twenty minutes before use. You can see where all this is leading. Paint that is at a reasonable temperature and sprayed onto a warm, dry job in a warm, dry atmosphere gives the best results. If you really need to paint on a cold or very damp morning, wrap a hot water bottle in a plastic shopping bag and leave it in the box with the job. All of this gives any sprayed paint the best possible start in life, with a good bond to the metal. You will have put in many hours to build your locomotive; don't skimp on this last stage.

Copper (and its alloys of brass, gunmetal and so on) is a 'slippery' metal. Indeed, it is sometimes used as a lubricant. This means that paint doesn't naturally want to bond to it. For a durable finish, we must give it some encouragement. The way to do this is to roughen it up at a microscopic level. This is normally done by etching, in the form of an etching primer paint. The alternative is a piece of the very finest grade of wet & dry paper, which does the job mechanically. The scratches are far too small to be visible, but they give something for the paint to grip to.

Remember that sprayed paint is extremely thin. The job may look the right colour but can easily be chipped. Build up with several applications; even six coats will be thinner than something you might put on with a brush. Leave plenty of time between coats. Also remember that just because paint is dry to the touch, it still won't have reached its proper hardness, which can take days to achieve. I advise patience, knowing full well that there have been times when I, like others, have lacked it.

MASKING OFF

You will be familiar with ordinary masking tape. It can sometimes be a bit brutal on our delicate paint finishes but it has its uses. You can lay a strip down on a piece of glass and then peel it off. This helps to 'de-tack' it and lessens the risk of pulling a previous coat of paint away. Whilst it is on that piece of glass, it can also be cut into fine strips with a scalpel and steel rule for more intricate masking tasks. Indeed, some shapes can be cut from it.

You can buy 'low-tack' masking tape especially formulated for modellers. Another useful option is a masking fluid that is painted on, subsequently peeling off as a fine skin. This might typically be useful around the polished cap of a chimney that you want to keep bright whilst spraying the rest of the assembly. For small parts that you want to keep bright, such as the shiny metal rings that surround cab window openings or a small brass whistle, you can carefully paint grease of some sort onto them. It is a bit messy but, given care, it does work.

It makes sense to spray, or paint, wheels fairly early on in proceedings. You could mask off the bright rims and bosses, but it is easier to paint the lot and then put a small wire cup in the bench drill and tickle the paint off where you don't want it. Hold the wheel on its axle and move it against the rotating cup. You will be surprised at how much fine control and accuracy you can get. For really small scales, use a tiny brush held in a Dremel tool.

Traditional model engineers in the larger scales advocated painting the spokes of the wheels and then doing the final turning in the lathe. I am not comfortable with this. Subjecting a paint finish to flying metal dust just doesn't seem right, but, of course, I bow to those for whom it works perfectly well.

TREATING METALS

Colour can be applied to metals in other ways than painting them. Steel can be 'blued' or 'browned' by going to a sports shop that sells guns and buying some bluing or browning solution. This gives good rust resistance and has a particularly nice patina. Copper and brass can be blackened chemically; I use Carrs' materials because they are available through model suppliers. If you have some brass shim or chain that is bright and shiny, heat it up to a dull red heat and then drop it in oil. This gives a nice dull finish.

APPLYING TOPCOATS

Previous remarks about spray painting also apply to the finishing coats. Many modellers will take advantage of the convenience of car paints, available in many colours. In specialist stores, the range may be even larger. But there are several things we need to bear in mind. It is important to remember that some of these have been cellulose-based paints. If sprayed over existing enamel-based paint layers, they can attack them and the resulting mess is horrible, with the result that there is nothing for it but to strip everything back to bare metal and start all over again . . . most disheartening. Fortunately, acrylic paints have become much more common for auto use now.

The other point relates more to aesthetics. Although there may apparently be a huge range of colours to choose from, many colours will be just too modern. We will look into this below. Fortunately, there is an alternative in the form of aerosols of paint formulated for railway modellers, with good interpretations of the original colours.

We will not dwell long upon the use of airbrushes, as it is too big a subject to deal with here. However, the use of them is a skill worth learning if you are going to build more than a couple of engines. The subject is well documented in books and DVDs. Airbrushing can produce superb results in experienced hands, but is not a passport to instant perfection in the hands of a beginner. If you master the subject, you will be set up for life. Should you decide you want to pursue this, I suggest that you buy an airbrush and a pressure source and become used to it whilst building the loco. A newly built steam engine is too precious to be your first test piece.

BRUSH PAINTING

Brush painting, on the other hand, has been with us for a long time. It is easy to do badly, but a bit of thought and planning equally makes it easy to do well. Oil-based paints – including those in small tins sold to modellers – respond well to thinning with the refined spirit sold as fuel for cigarette lighters. However, conventional thinners work well. All the usual rules apply. Stir the paint thoroughly and when you have finished, stir it again. Don't begrudge buying new paint for a newly completed engine. Use good soft brushes. Don't try to cover with one thick coat, but also be patient and leave plenty of drying time between two or three thinner coats. You will have read all this countless times before so I won't expand on these points. As soon as you have finished applying paint, cover the job at once. A cheap plastic propagator is useful. I use a defunct microwave oven as a spray booth, but it is also just right for popping newly brush-painted models in and closing the door.

However, there are a couple of things to add. Modellers' oil-based paints have slightly different characteristics: which vary from colour to colour, with each even smelling slightly different. It makes sense, therefore, to open a new colour and try a few experimental passes on a piece of scrap metal, in order to get the feel of how it behaves, before committing brush to model. For example, you will discover that some yellows need to be applied with more coats than, say, greens. Even after you have applied several coats, there can still be variations. If you want the red on buffer beams to 'sing', apply a couple of thin coats of white first.

Many years ago we used to bake enamel paints and lacquer finishes. The job would be left for a day to become touch dry, then would be put in a gas oven at a low setting for a few hours. Thereafter, it was put away in a dry, dust-proof area for a week. It seemed to impart additional strength to the job, but I could never quite be sure just how valuable it was. It seemed good practice at the time – although of course the ability to do this usually depended on the harmony of one's domestic arrangements.

Don't try to cover large areas, like the side of a tank, with a small brush. I find it a good rule of thumb to think of the size of brush I should be using and then choose one slightly bigger. It seems to help the paint to flow more smoothly. Aim to keep flat surfaces horizontal so as to avoid the risk of runs. You can't do this on round things like boilers, so in that case it is just a case of keeping the paint thin and being patient enough to apply several coats. It does help if the metal is warm. Make up little hand rests so that your wrist is supported as you paint, rather than trying to hold a brush steady at the end of an extended arm.

All paints have two drying times – touch dry and hard. Don't confuse the two. Although new paint may feel dry to the touch, it can still be soft enough to be spoilt by handling. Be patient and leave the job until tomorrow – or next week if possible, whilst you get on with something else.

Acrylic paints are becoming more widely used. They have low odour in comparison to oil-based paints and the brushes clean out in water. Methylated spirit makes a good thinner for a smoother finish, instead of the more usual water. Artists' acrylics are a different formula to modeller's paints. Until you know the medium well, stick to those sold in model shops.

Heat-resistant paints are available for such things as customizing cars and motorcycles. In theory, such things should be perfect for small steam engines, but unless preparation is scrupulously done, there is a strong risk of paint failure – it bubbles up and falls off. When applied with a brush, it can also be a bit crude. So it is one of those things that should work, but which often gives a high failure rate. You may come across people who have had the greater success that has eluded me over the years. I would mention one useful matt black high temperature paint, which is sold for respraying black barbecues. It appears to cope with heat quite well, although WD-40 in particular seems to want to wash it away.

You can buy brushing cellulose paints, together with appropriate thinners, but there is a feeling that they offer the worst of all worlds. Yes, they will provide a good finish with care, but there are problems with smell and the fact that cellulose can attack other paints. However, cellulose paint seems to have died out at the time of going to press.

THE AESTHETICS OF PAINTING

I am fully aware that this is dangerous ground. Some modellers are concerned with the engineering attributes of a locomotive and happily say that they don't really care how it is painted. Tastes vary. There are those for whom an engine is a superb shiny object in its own right and anything that detracts from this should be avoided. An alternative view is that a high gloss doesn't scale down accurately with distance, so satin-finished paints look more realistic. Taking this a stage further are those who prefer 'scale colour' so that their models look like the real thing shrunk down.

With the usual disclaimer, I draw your attention to Phoenix Precision Paints, which contain a large range of railway colours that seem to be accurate to the prototype. Phoenix manufacture paints in and and gloss finishes and in aerosol form too, and they can also supply low-tack masking tape, etching primers and many other products.

There are various painting dodges to improve the appearance of a locomotive. If your metalwork has produced an occasional poor joint, paint it with satin finish instead of gloss. It was common practice for the insides of cabs to be painted a lighter colour. If you do this, the cab will suddenly seem to have an inside and an outside like the prototype, instead of being a collection of painted metal sheets. Although not always correct, picking out brake levers or regulators in red can give things a lift. I have said elsewhere that I think that a wooden planked cab floor adds a certain something. Try to be really observant. Most buffers have stocks that are red; occasionally they are black. Buffer heads and stems are left as shiny metal. The sides of tyres on wheels are usually left unpainted. There are all sorts of little errors that one sees from time to time that are entirely due to a lack of observation – and I speak as someone who has made such mistakes.

BUMPER PAINTS

A useful resource is the small range of plastic bumper paints for cars. Halfords, for example, offer it in two shades of grey and a black. It is a dull finish that generally sprays onto a model beautifully, without seeming to attack other types of paint.

Note that piles of coal are *not* matt black; they tend to be erratically shiny. Headlamps for engines have bodies that are painted red, white or black. If you are using cast metal lamps that have the lens moulded in, please don't paint them white, silver or red. In daylight, the colour you actually see may well be what I would describe as a dark grey/brown/green, or perhaps a very deep shiny maroon for a red light. Next time you are standing by a prototype engine, have a look for yourself and make your own judgement.

And now we move into really subjective territory. In the absence of painting a loco in a specific

Case Notes:

Kiso

Kiso represents one of the variety of Baldwin engines, in this case exported to the Kiso Forest Railway in Japan. To my eyes, they were dainty little things: somehow attractive in their awkwardness. You can see that they represent American practice with bar frames, thus giving a slightly airy look below the boiler. By now, you will have come to appreciate that most of the construction techniques tend to remain the same, no matter what the prototype, but we need to think about how to tackle several unusual features.

I would suggest that the bar frames are fabricated by pieces of thick brass bar – perhaps 4mm thick for G scale. The two 'main beams' would be silver soldered to axle box rectangles in a jig. The valve gear to the outside cylinders would be derived from slip eccentrics, between the frames, running off the back axle, meaning that rocker arms would be needed to transmit this motion to the outside. This would take care of a pair of commercial cylinders.

At a deeper level, we could flop those cylinders over on their side so that the valve chests were between the frames, although that could mean 'interrupting' those main bars to make room. This starts to become, as Sir Arthur Conan Doyle said, 'a three-pipe problem'. An alternative would be to make it the subject of a conversion from an Accucraft Ruby kit, as discussed in the Nielson box tank described in Chapter 5.

That blued boiler finish is distinctive. As the boiler is made of copper, that can't be given a suitable treatment at home, so we need a thin outer wrapper. This could be made of mild steel, which we roll and then blue with gun blue. An alternative is to buy a piece of steel that has been commercially electroplated. You may also be able to get such a finish in aluminium as well. It is worth the effort of tracking it down to be able to get that excellent effect properly. A more restrained treatment for a boiler wrapper like this is known as 'Russian Iron'. Again, you might feel it worth investigating.

livery, there seem to be some aesthetic practices that have grown up during the development of the steam engine. The rounded boiler shapes and odd angles seem to respond better to warmer, darker colours. It is no accident that variations of a dark green have been the most commonly used colour, other than black. I put forward the personal view that the correct GWR later green works so well because it is a rich, soft colour, leaning towards olive. Polished metal seems to enjoy being in its company. I am very much a Southern man but could never reconcile myself to Malachite Green except, possibly, on the flat surfaces of Bullied Pacifics.

Discussions of such subjects have gone on for many years. They seem to generate more heat than light. In the absence of any other constraints, I would suggest that you stick to traditional darker colours for your first, freelance, models. Could I particularly put in a plea that you avoid the basic 'Meccano-type' colours so beloved by some commercial manufacturers. They can make models seem toy-like, when this can so easily be avoided. They may use these colours for sound commercial reasons; we, on the other hand, can take the time to be more discerning.

SCALE COLOUR AND WEATHERING

There are two schools of thought here. The first is that a model steam engine is a working piece of machinery in its own right and, as such, it should be painted in the correct colours and usually in gloss paint too. The end result is attractive – and the loco is often more sellable when it gleams.

The alternative is that a model should be a believable representation of the real thing, seen at a distance. And here we encounter the fact that 'scale colour' kicks in at a much closer distance than distant hills do when they turn blue. Black starts to turn to dark grey in less than a hundred yards. Alas, our brain automatically compensates for that. It tells us it is black, even though it is actually seeing dark grey. Similarly, white turns to pale grey. Colours soften. An artist has to learn to paint what he sees, not what he knows.

Very often, an architectural model, although simplified, has a rightness about it because it is painted in restrained colours. It is like the colour contrast on a television being turned down and slightly lightened. Instead of seeing a well-built model, we are seeing the real thing at a distance. But, again, I stress that this is just a way of looking at things and may not be to your taste at all.

Weathering, on the other hand, is a much more precise matter and is based on observation. It does *not* consist of splashing matt black paint over everything. There is a faint grime colour that is best applied with an airbrush. A faint haze of a medium dark grey lets colours underneath come through. Down below the running board the grime should have a slight brown tinge to it. This is caused by the presence of dust, generated when the brakes are applied. Do not overdo rust streaks or white deposits of lime in water or steam (in the case of the latter try painting a little diluted metal polish on the area required and let it dry). Under the running boards, weathering mostly consists of a faint brown haze, produced by iron dust being thrown around when brakes are applied.

There is a whole subgroup of weathering that uses chalks and pastels. I had most success with crushing them into powder and blowing that onto the model with a drinking straw. When satisfied, a few whiffs of matt varnish spray are applied to fix the dust in place. I have used this technique on steam engines. It can look impressive when properly done. That said, I have reservations about blowing dust around working parts; nor am I too happy about using varnishes on live steam engines. They can go yellow and dull in regular proximity to heat.

If you really want to go into the subject artistically consider using a dark grey – with or without a brown tinge – instead of black. I have very fond memories of sitting by the paint shop of Eastleigh Works on occasions, looking at a newly turned-out engine or two. The black looked almost as though it had a blue tinge to it – reflecting the sky, no doubt – and appeared quite unnatural to my eyes, which were accustomed to normal working engines. Great Western buffer beam red had a particularly orange tinge to it and, when new, it was very vibrant – almost bordering on dayglo. But, given the limitations of paint quality, this soon dulled to a brick-red colour.

This is all rather an ephemeral subject, given to endless argument and discussion. Forests have been cut down to make the paper containing the discussion regarding what the infamous LBSC 'Improved Engine Green' really looked like – in both new and used condition. The Caledonian

Railway used a magnificent blue, but which varied enormously in shade, depending on how much white base was added as an economy measure at a particular works. Not always documented are the ways in which, for example, German and French colour palettes are different to British ones. Old oil-based paints had qualities of colour that were different to modern synthetic paints.

These are fascinating subjects for discussion, but perhaps we can leave their niceties for times when we are further along our loco-building career. A company like Phoenix Precision Paints can be relied upon to produce accurate paints for our needs. Some people take colour extremely seriously, but perhaps we shouldn't navel-gaze too intently. There will be loco-builders who have no feeling for accurate colour. Indeed, history has thrown up some good modellers who are unfortunately colour-blind, but their models will give them as much pleasure as anyone else's – and will work just as well.

All weathering should be so restrained that you don't realize it has been done. Not many people like to weather their precious steam engines, but these are the sort of factors to bear in mind if you do. A properly weathered engine is *not* the same thing as a new engine that has been left to become scruffy with age, atlhough even here there are grounds for argument. There is a school of thought that likes the patina and distressing that a well-used model acquires.

LINING

For every beautifully lined locomotive that you will see, there will be at least one that looks awful. The trouble is that the slightest flaw is instantly obvious. I could give you the advice to buy an expensive bow pen or lining tool and to become proficient in using it. That is the theoretical perfect answer, but not everyone is so gifted. I am going to suggest that the first option is to leave off the lining. No lining is better than wobbly lining. Another option is to use one of the newer lining pens that feature a tiny roller. They are kinder to beginners as the paint flows more evenly. It is possible for a beginner to rule a clear straight line after a little practice. By using a piece of masking tape at the beginning and end of the line, it also starts and stops in the right place. One has to be patient and draw a couple of lines and then leave everything to dry before doing more, but this method can produce a reliable neatness for simple lining.

Another alternative is to use lining tapes. With care and patience, they can be applied neatly. Some people are naturally neater than others. Crooked tapes look 'wussanawful'! There are some people who just don't have an eye for such things – indeed, they may not be aware of true 'squareness' – and are happily blind to the faults. I stress the word 'happily'. That is important. If they are happy with what they do, then we wish them well and share their pleasure. But if you cannot apply tapes really well, again, perhaps they are better left off. Some tapes have been known to lift off hot metal, particularly with spirit-fired engines, where an occasional stray lick of flame can attack them.

Likewise, the physical demands of a hot oily engine, particularly a pot boiler, may be too much for transfers. So that leaves the option of getting the engine professionally lined. This might seem a luxury, but, considering how much you have saved over buying a ready-made engine, perhaps it is one you can treat yourself to. In some scale groups, there will be someone who does the job reasonably and who will take in engines at one show and have them ready for the next. In the 16mm arena, LightLine seems to be highly regarded. Again, I offer my usual disclaimer.

BRIGHTWORK

Boiler bands can be polished brass or copper. This is not always correct to the prototype but looks attractive. Bespoke etched brass name and number plates are so freely available that it is hardly worth making your own. In years past I have used Letraset and ferric chloride to produce them myself, but I wouldn't bother these days. However, we now have the delights of A4 paper that is a printable etch resist (stocked at Maplin Electronics or similar). Artwork is built up in the computer to your heart's content. It makes sense to fill a

A selection of locomotives that have been professionally lined by LightLine. It shows how much good lining can 'lift' a locomotive. (Photos: courtesy of Geoff Munday)

sheet with printing before applying it. It is then ironed on to printed circuit board and etched in the usual way.

My personal feeling is that it is not worth putting a sealing coat of varnish or lacquer over polished metal to maintain its shine, particularly with live steam engines. Yes, it will stay shiny for longer than unprotected metal, but it will be much harder to clean up when it eventually dulls.

If you want a simple dodge for painting a boiler band that is lined colour–black–colour, paint the body colour all over a piece of thin metal sheet (nickel silver is a good choice). Then, with a lining pen, simply rule a few lines with black paint, with spaces between them. Leave this to harden off, then, with a steel straight edge and a scalpel, scribe lines a short distance either side of the black line. These scribing lines will act as a guide

The weary weathered look. This Mamod re-engineered conversion was built by Eric Lloyd and Colin Binnie.

to let you cut out the bands accurately with sharp scissors.

Despite the different approaches and opinions about painting and lining, the same basic rules hold good. Just because it is the last stage in building an engine, don't rush it. Be as scrupulous as possible; try to get as good a finish as you can. There is a reworking of an old saying: 'A coat of paint hides a multitude of tins.' It is the paint finish that we actually see.

CONCLUSIONS

Let us finally draw the threads of this book together. We have looked at metalworking techniques, but decided that building a small live steam engine is possible without extensive machining. While the subject can be taken to greater depths with your newly gathered experience, there is nothing that says you have to become a skilled engineer. Work at whatever level you are happy with. Most loco-building is still about the basics of getting a straight line and a right angle. It may be that you have no ambition to be a loco-builder for its own sake, but are simply looking to build engines as part of your overall garden railway plans, in which case there is no need to let yourself be talked into going beyond your own preferred level of working.

You will also have noticed that this book has used projects to describe different ways of building. The aim has been to show that techniques can be applied to any design you choose – it is just a case of studying the drawings and working through any problems they throw up and then starting to cut the metal. For example, a body built around Roundhouse cylinders and a boiler can be any shape you choose. You have some basic dimensions to start with, and your design would need to be appropriate to those, but the technology remains the same.

You can also mix and match techniques. If you want to build a Gauge 1 LNWR coal tank with a single Roundhouse cylinder between the frames and a gas-fired boiler, it can be done. The chances are that you will evolve ways of working that will suit your needs best, although you should try not to get so locked into them that you ignore everything else. Alternatively, you may be the sort of person with an ambition to advance your skill levels and expand your workshop facilities. But whatever your long-term goals, never forget that this is something we do for pleasure. Of course, there will be occasional setbacks. There is nothing more frustrating that making a component and then finding that it is slightly too short. It happens. Just get on and make another one. There is a way round every problem.

I wish you the same pleasure and interest that this subject has given to me over many years. It is done: our explorations together are complete.

APPENDIX I

Index of Loco Designs

Here is a list of the designs that appear in this book. Some are detailed drawings, whereas others are merely rough sketches or outlines. The principal gauge is indicated first, but possible alternatives are suggested. Where I use the term G scale, this tends to imply compatibility with LGB, at 1/22.5 scale. The term 'Gauge 1' refers to the British 10mm to the foot scale. But these suggestions can be flexible. If you want to build in 1:24 or 1:29 scale, then the calculator and computer will help you to change the sizes to suit.

Anton	LGB compatible; a powerful loco that will handle sharp curves	G
BAGRS	Brief sketch of a typical US-style 'speeder' project	G
Bardot	Typical French tank engine built over a Roundhouse chassis/boiler	SM32 / SM45
Brecon	Early crude sketch of a very basic tank engine	SM32 / All
Bunty	Diagram only of a simple vertical boiler loco	SM32 / All
Cranefly	Early crude sketch of a crane loco	SM32 / All
Dacre	Simple loco, and variations, designed around commercial parts	SM32 / SM45
Dante	5in gauge wing tank; a compact loco, suitable for 2.5in gauge or as a 16mm scale loco	
Dempsey	Early box tank, based on Ruby kit	G1 / SM45
Denys	Glyn Valley Tram body to drop on various chassis	SM32 / SM45
Gnat	Rough sketch of a very simple loco with a cut-down body	SM32 / Any
Greyhound	General outline of the Drummond T9 in original form	Any / SG
Java	Generic East Indies sugarcane loco built on Roundhouse chassis/boiler	SM32 / SM45
Java Steam Tram	Example of building a simple loco around a stationary engine	SM32 / Any
Klaus	Basic design for a working steam rack locomotive	G / SM32
Lagos	Semi-scale model of a Nigerian steam tram, using 16mm components	⅞in scale
La Petite	Very small enclosed cab tank engine	Any N/G
Lewin	Outline for a simple haycock firebox loco	Any
Lyonesse	Geared mineral loco	SM32 / Any
Margaret	General outline of Fox Walker saddle tank	Any SG
Motor Tank	LNWR Motor Tank designed by Harold Denyer	G1

Prince	Thoughts on a geared loco, suitable for flanged track waggonway	Any
Sussex	Traction engine type	G1 / G / 2.5in
Terrier	Conventional standard gauge loco	G1 / 2.5in
Vulcanette	Dainty standard gauge loco	G1 / 2.5in
Wirral	Beyer Peacock type 2-4-0 tank engine	G1 / 2.5in

APPENDIX II

The Aster Connection

In a book that is aimed at beginners, it would be remiss of me not to include a reference to the steam locomotive kits of Aster Hobbies (UK) LLP. Here, pretty well all the machining and painting has been splendidly executed to a high standard. Putting one of these kits together is mostly straightforward assembly work. At first glance, a kit price can seem high but, given the amount of work that has gone into producing quality components, it isn't. You are paying for a lot of time and skill – and I really do mean a lot. Moreover, you will not need to put together a well-equipped workshop to arrive at the happy state of having built a steam engine.

Aster has a rolling programme of producing kits for a while and then discontinuing them, to be replaced by new items. Nearly all of their products relate to standard gauge prototypes in Gauge 1.

They are *not* instant plastic kits. The builder is called on to exercise care and some thought. There is no machining to be done, in the normal course of events, although some tweaking and adjustment may be needed. But what you do get is an end product that looks magnificent, with factory paintwork and lining.

Some Aster designs seem to take a positive delight in the challenge of producing complex engineering. They can be desirable objects in their own right – real technical achievements. But it makes sense to cut your teeth on something simpler first. It helps you to get a 'feel' for the way things are done and the way the manufacturer thinks things through. In Chapter 4 of the book, we discussed building an Accucraft Ruby kit as a way of getting introduced to the feel and shape of all the bits in a small live steam engine. Building an Aster kit takes this idea further along. I would urge that you don't plunge headfirst into one of the more complex larger locomotives, but cut your teeth on a simpler model first.

If you embark on one of these kits, you will soon become aware of an Aster 'housestyle'. The components are made in a factory environment. Parts might be nicely plated. Cylinders are usually operated by piston valves. These call for very accurate machining and so we may well avoid making them in our home workshop, but they work well when precision engineered. So some Aster components may look a bit different to what has gone before in

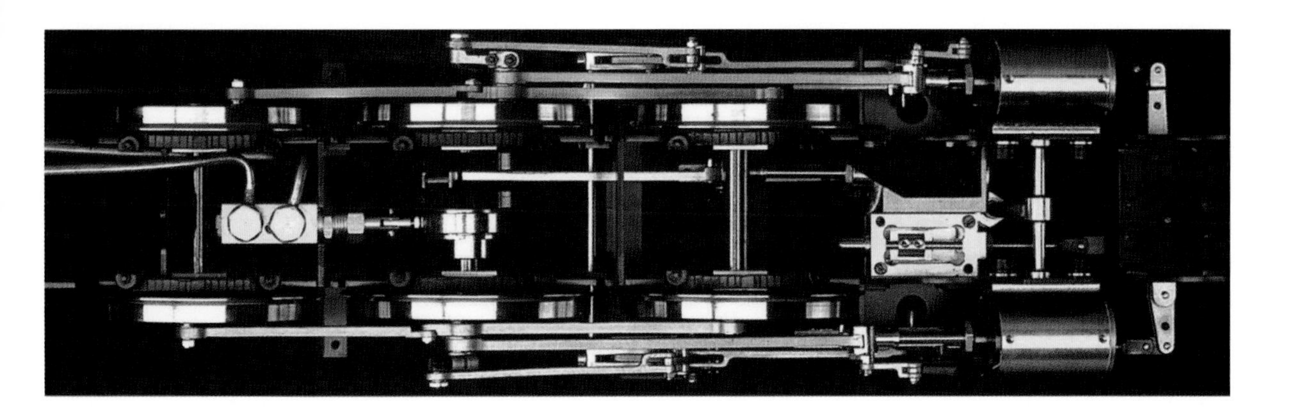

this book, but do exactly the same job and you will soon be familiar with them.

The models are mostly internally fired. Particularly with regard to the larger models, they are seen at their best on high-level tracks with large sweeping curves and nicely laid track. This, I suggest, is their true home. Again, I offer the inevitable disclaimer, but I have found Aster UK to be particularly helpful.

APPENDIX III

Testing and Insurance

This section will, of necessity, be fairly formal, and slightly daunting at first sight. There are legal and insurance considerations at work here. What you do in the privacy of your own workshop is up to you, but when you are at meetings or where the public are invited, you should follow the wise words. They can all sound rather overwhelming. The booklet produced via the Association of 16mm Narrow Gauge Modellers, in conjunction with the British Model Engineering Liaison Group, runs to twenty pages. However, it all comes down to 'get your boiler tested and obtain a certificate'.

There is a code of practice, agreed by the principal provider of insurance to the model engineering hobby. Boilers of less than 3 bar litres (which includes most of what we are interested in) are exempt. But it is strongly recommended that any small boilers, fitted with a safety valve and a pressure gauge, are tested every twelve months, and that the pipework is checked at the same time.

1 bar = 14.5038 psi	10 psi = 0.689 bar
1 litre = 0.22 gallons / 1.76 pints	
1 pint = 0.568 litre	1 gallon = 4.544 litres

Note that, because of the different issues of construction arising, stainless steel boilers are not covered under the particular code we adhere to normally. This is probably academic because it is very unlikely that a beginner would build a boiler in this material.

If the boiler being tested is not a recognized existing design, the builder should satisfy the inspector that the design and materials offer adequate strength.

A boiler will be tested hydraulically to twice the normal working steam pressure. A steam pressure test of 150 per cent may be carried out.

These are just a few points extracted from a much longer, more precise, document. A copy of this should be available from your particular scale/gauge organization. But don't get things out of perspective. In the end, what it comes down to is that any boiler you build ought be tested and certified for insurance purposes. It is usually a matter of a short time with an approved tester, often at meets. The Association of 16mm Narrow Gauge Modellers, for example, has a number of approved boiler testers, in different areas. It is worth getting a new boiler checked over anyway, just for your own peace of mind. Commercial boilers usually come with a certificate anyway.

Having stated the facts of boiler testing, it is worth noting that boilers below a certain size do not currently need a boiler certificate – at the time of writing. But it still makes sense to get a boiler checked out.

There is some debate as to whether gas tanks should be tested. They are usually so small physically that it seems like a bit of overkill. Some loco manufacturers provide a certificate; some don't. At the time of writing, they do not seem to be a requirement, but if in any doubt contact your association for advice. As a last resort, contact an appropriate insurance provider but be prepared for conflicting advice. They may not have the mindset to consider the needs of our tiny models, being geared up for dealing with much larger boilers.

Bear in mind that this subject is somewhat fluid. Keep an eye on material coming from your particular association about new amendments.

APPENDIX IV

Adhesives

Adhesives don't play a major part in the construction of live steam locomotives, but they have their place. Sometimes it is an unusual place.

The big rule to remember, right at the start, is that the cyano group of adhesives, such as 'superglue', tend to fail under heat. So it is not a scrap of use using ordinary superglue for fixing wheels to axles. We have to remember a particular principle here. Live steam engines in our small scales get hotter than big passenger-hauling locomotives. This is because everything is closer to the source of heat and because there are not big chunks of metal to act as heat sinks. For example, you can pick up a hot 5in gauge engine by its buffer beams, but the buffers of its smaller brethren can burn skin.

We can make use of the fact that cyano adhesives can fail under heat. Parts can be temporarily glued together and then drilled through or sawn. After the job is done, a whiff from the blowlamp will free them. It may not be best engineering practice, but it is worth mentally filing this trick away.

There are a whole group of engineering adhesives that are related – Studlock, Nutlock and all sorts of things-lock. They repay study and research into the subject. They do what their name suggests and are usually anaerobic, that is, they set in an absence of air. Ordinary cyano can also be thickened out with 'Rocket' powder, a commercial filler, to form a fillet. A substitute for this is washing soda. But all the time, keep in the back of your mind that some of these adhesives are unreliable when stressed by heat. Always read the small print of the instructions. Or if in doubt . . . use a rivet instead.

Bear in mind that you can use a liquid gasket. This is usually an orange-coloured paste that is smeared on the surfaces. They are then bought together and clamped tightly for a day or so. Be warned that this stuff likes to fill up ports and fine holes if you are not careful to wipe away all surplus.

Slow-setting Araldite, properly applied, will stand the heat of a small steam engine. I am still loathe to trust it for anything that is subject to shearing forces, but, for example, if you want to give your cab a floor of scale wood planks, then slow-setting Araldite will be fine. As a rule of thumb, don't trust any sort of rapid-setting Araldite or similar. Be patient. If you need to hold two parts together whilst the glue is curing, use a couple of dots of superglue on a clean part of the job. This may fail under heat but only in a very small area. The rest will be held together well.

The general-purpose adhesives such as Bostik have limited applications, but still have their uses, for example gluing coal grains into a tender. They usually have a strong smell and need to be used in a well-ventilated area. This also applies to many impact adhesives.

There are low-odour adhesives that have fewer active ingredients and are kinder to use around the home. I have found them to be less effective, although they still have a role to play in less critical joints.

For making tram bodies or wooden tenders, there are a variety of good woodworking glues. I generally keep a small quantity of waterproof PVA in an open container for some hours before use. This makes it more sticky – or 'hi-grab' in the current vernacular. For really powerful wood glues, seek out information from a boat chandlers or investigate online. I have a particular affection for Aerolite 306. Being pragmatic, I have found that, where possible, all adhesives work better when they are reinforced with a pin, a nut and bolt, or a screw. Sometimes it is useful to drill a hole to accept part of a cocktail stick to add mechanical reinforcement to a glued joint. I would also echo the obvious advice – read the label or leaflet!

You are unlikely to need plastic adhesives anywhere near a hot steam engine.

Finally a word about buying budget brands. I have found that if a store brand sells bottles of a particular adhesive, such as cyanoacrylate, then that is what you get and it works just as well as a name brand.

APPENDIX V

Metals

Throughout the book we have spoken of different types of metal in fairly loose terms. Here is a more definitive guide, listed alphabetically, which pulls all the strands together in one place.

ALUMINIUM

This presents itself in pure form or in a huge variety of alloys. These conveniently fall into two groups: those that can be hardened with heat and those that can't. The most well-known alloy is Duralumin. It is very strong for its weight, but, for model engineers, it has the drawback of being difficult to braze easily. In industry, argon-arc welding is used. With age, there is a risk of aluminium developing a surface of powdery oxide unless protected.

BRASS

Brass is an alloy of copper and zinc and comes in different varieties.

- Brazing brass is used for brazing steel articles. Generally 75 per cent copper and 25 per cent zinc.
- Deep drawing brass is intended for press tools and is more ductile. 70 per cent copper, 30 per cent zinc.
- Free-cutting brass is good for turning. 85.5 per cent copper, 10 per cent zinc and 1.5 per cent lead.
- Screw brass is inferior for most of our purposes. 78 per cent copper, 16 per cent zinc, 4.5 per cent tin, 1.5 per cent lead.
- Tube brass is harder than normal brass. Used as name suggests. 65 per cent copper, 35 per cent zinc.
- Clock brass – good for gears. 62.5 per cent copper, 35.75 per cent zinc, 1.75 per cent lead.

Brass tube could be used for boilers up to 25psi, but it is generally preferable to use copper.

Brass sheet usually comes in three different states – hard, half-hard and soft. Hard is the most useful to us for platework and general use.

Cultivate your own hoard of little brass bits in a box somewhere. The pins from electrical plugs, small brass ornaments and the like spring to mind. Once you start looking, it is amazing how many redundant objects will yield up small brass items. In days gone by, we used to find many uses for old bullet cases, but these are less accessible these days!

CAST IRON

This is a soft, granular material, much used for machining components. Swarf comes off in the form of very fine grey chips/powder. In castings, there is a tougher outer skin. It will rust much more slowly than mild steel. Somewhat brittle in thin section in particular (as we who have dropped a newly finished turned driving wheel onto a hard floor will testify). Castings sold for model steam engines often have a small amount of nickel added. In past times, I have found old sash window weights were a useful source of this material (although it is always worth keeping an eye open for an occasional blow hole).

COPPER

Copper is a very ductile metal; it can be beaten into shapes easily. If you watch a good coppersmith at work you will see that he almost gets the metal to flow. Thus a dome can be produced from a flat sheet of material. It will probably be beyond the beginner to produce results like that, although flanged boiler ends can be beaten over formers by a newcomer, given practice. But it is with tubes that we will find most use for this material. Boiler shells and inner tubes should be cut from seamless, solid-drawn copper. Copper sheet, bought for boiler-making, will probably be very pure (over 99 per cent). Beryllium-copper develops enormous tensile strength when heat-treated but it also has good elastic properties and fatigue resistance. Suitable for high quality springs.

GUNMETAL AND BRONZE

- Gunmetal is typically 88 per cent copper, 10 per cent tin and 2 per cent zinc. Machines nicely.
- Phosphor-bronze is used for bearing surfaces. Good for boiler pumps and fittings generally.
- Lead-bronze is not a true alloy. Used for very heavy-duty bearings.
- Manganese-bronze is *very* tough and corrosion-resistant. Excellent for boiler fittings.

NICKEL SILVER

This is an alloy of 55 per cent copper, 25 per cent nickel and 20 per cent zinc. It is a lovely material for platework and it retains paint much better than brass.

MILD STEEL

The bedrock of model engineering. Available in a huge variety of sections, rod, sheet and bar. Very prone to rust unless protected. For turning round section well, look for 'best commercial quality' such as Ledloy or similar. This contains a small percentage of lead, which helps to put a good finish on a turned job. For axles, think about buying round ground mild steel. Be aware that steel comes in a variety of qualities. Do *not* try to economize by using offcuts of concrete reinforcing bars. I am a keen advocate of making use of recycled materials where possible, but for specific loco-building tasks, don't begrudge buying from a specific stockist for model engineering. Flat spring steel is a variant that is particularly suitable for laminated springs.

SILVER STEEL

A high carbon steel with no silver content. More difficult to machine than mild steel. Its specific application is for hard-wearing jobs such as crankpins, after hardening. Bar stock is useful for making motion work, but harder to work with.

STAINLESS STEEL

There is a variety of alloys that come under this heading – some are highly resistant to corrosion, some less so. But most are very tough to work with. The alloy type 18/8 is commonly used for model engineering practice and is typically used for valve and piston rods.

In the lathe, make sure your cutting tool is sharp and start the turning off with a deep cut. Such metals will rub instead of cut, given half a chance. This dulls the tool and work-hardens the metal. A useful dodge: a good source of thin, flat section is the cutlery drawer in the kitchen. But don't get caught by your family! There is also an alloy called 'stainless iron' (KE.40A), which is easier to machine than stainless steel.

TINPLATE

This isn't a metal or alloy as such. It is mild steel sheet, coated in tin. But it is worth mentioning as an excellent sheet material. It solders easily and paint adheres to it well. It needs paint protection to prevent long-term rusting. A sadly neglected material these days.

WHITE METAL

This is a generic term that covers a large family of low melting point soft alloys. The most common application is in white metal castings. They have low physical strength so long thin sections are easily broken. They will stand a moderate heat but can melt in the sad event of a locomotive conflagration or if an engine is allowed to boil dry. If you want to make your own castings, use proper casting metal from a model suppliers. They will also provide you with suitable mould-making materials.

Not quite in this category are some of the alloys, like Mazac, which old model wheels used to be made of. They looked fine when new, but, after several decades, they would crumble away. And there's nothing that can be done to stop this. If you take on an old engine for restoration, you may find that you will have to replace the wheels with cast iron ones, although it is a problem that has largely died out now. However, whilst we are on this topic, could I suggest that you remain suspicious of alloy wheels for another reason. There are some compounds that have a greater coefficient of expansion than iron or steel. When an engine gets hot, the wheels become loose on their axles. So stick to brass, steel or cast iron. Also avoid white metal wheels that may some day cross your path. They are usable for static models or light battery locos that see very occasional use, but not for live steam engines.

COLOUR CODING

To identify stored metals quickly, it is worthwhile colour coding them. The commonly used designations are:

Metal	Colour
* Black mild steel	No colour
* Bright drawn mild steel	Aluminium colour
* Brass	Dark red
* Stainless steels	Bright red
* Tool steels	Blue
* Gunmetal/bronze	Green
* Cast iron	Black

APPENDIX VI

Other Commercial RTR Steam Engines and Parts

In this book, you will have seen that there is scant reference to using Mamod steam railway engines as a basis for conversions. This is intentional. When they are good, they are excellent. But the name went through a rapid turnover of owning companies, some of whom did not understand the need for the quality control that reliable steam engines demand. Some locomotives, and especially kits, were so badly made as to need considerable expertise in order to get them to run properly (indeed, some could not run at all!). If you have a wish to build an engine out of a simple Mamod, make sure that you get one that works properly in the first place. They can provide a basis for a simple engine with a few tweaks, for instance lap all flat working surfaces to improve on the finish and flatness of them. Deryck Goodall, who built the tram engine on page 88 from Mamod components, became the master of Mamod mechanics and devised many simple upgrades, such as extending the blunt ends of the semicircular canals in the reversing block with a file to a slightly longer point at each end. This gives a slightly better control over steam admission. I have enjoyed building a selection of conversions over the years – usually bringing them up to full 16mm proportions. But, and this is very much a personal view, I have never felt satisfied with their performance compared to a more conventional model.

There still seem to be plenty of unfinished home-made steam engines available, in 2.5in gauge in particular. Often they are based on old Bonds castings and drawings. It can be tempting to buy a part-finished project. It can seem quite cheap for what you get and there is the appeal of some instant hardware in your hands. But there are two things to worry about. Firstly, they can call for good model engineering abilities to complete them; something you may not have yet. Secondly, it is not unknown for a project to be sold on because there has been some serious hidden flaw in construction that each new successive owner has discovered. An example might be where the frames are a few millimetres too far apart. It is hardly noticeable on first inspection, but it means that the wheels will be over-gauge and cannot run on 2.5in gauge track unless the chassis is stripped right back to the raw parts and new ones made. Inviting though these part-builts may seem, they may not be for a beginner. However, at a later stage when you have got some experience, they can be a useful source of inexpensive wheels and cylinders, even if you don't plan to take the original model to completion.

I would draw your attention to the little Regner engines – Conrad, Willi and the like. They appear as slightly toy-like engines – rather shiny and powered by an oscillating engine. But appearances are deceptive. They are beautifully engineered and run slowly and well for long periods. They are a joy to operate. Moreover, for a complete beginner they make the perfect working parts to go under a home-made bodyshell, such as a tram engine. With hand tools only, you can produce your own nice-looking engine, which runs delightfully.

APPENDIX VII

Operating Small Steam Locomotives

This subject has been well-documented elsewhere, but I would like to draw a beginner's attention to a few particular points.

- Always use distilled water unless you are one of the fortunate few that live in an area of outstanding water softness.
- Never use anything other than proper grade light steam oil in the lubricator of a locomotive. In particular, *never* be tempted to use an automotive oil. It is not capable of standing prolonged heat and may well break down and carbonize, eventually blocking steam pipes altogether.
- Avoid cheap makes of butane gas cartridges. They are often 'dirty' and specks of dirt will be continually blocking your jets. The 'bush telegraph' will tell you which makes should be avoided. A tiny piece of kitchen towel, loosely crumpled, can be inserted in a union somewhere between the gas tank and the burner. This acts as a filter which lets the gas pass but intercepts the little specks of gunge that a dirty butane can produce.
- There has been much argument about different types of methylated spirit and its hazards. In particular, there is a view that reconstituted or industrial meths contains a lot of water and is less effective. I have never encountered the problem, but maybe I have just been lucky. Overseas readers will find that 'alcohol' (methanol) is the normal substance to burn. It is a clear liquid – and the wise loco man will add a drop of food colouring dye to it, so that a bottle of it doesn't get mixed up with his water supply.
- A light grade machine oil is fine for external lubrication and can be applied freely. If the underside of the engine gets a build-up of gunge from this, mixed with general dirt, then a spray of WD-40 will help clear it all away.
- At the end of a run, whilst the engine is hot, give it a wipe down with the traditional oily rag, perhaps aided by a few drops of oil and paraffin mixed together.
- As soon as a fire is extinguished, *always shut the regulator*. A cooling boiler creates a vacuum and will suck oil from the lubricator and pipes back into itself. You want to avoid oil in a boiler at all costs, because it acts as an insulating film, thus reducing the heating capacity of the fire.

APPENDIX VIII

Tables

Here are two tables that will be of immediate use as you study this book. There are many more to be had and I refer you to the Bibliography to source these. If you have any model-engineer friends, most of them will already have exhaustive tables on their shelves.

1) Gauge Number and Letter Size Drills + Equivalents

Gauge No. or Letter	Decimal Equivalent	Metric Equiv.		Gauge No. or Letter	Decimal Equivalent	Metric Equiv.		Gauge No. or Letter	Decimal Equivalent	Metric equiv.		Gauge No. or Letter	Decimal Equivalent	Metric Equiv.	
		B.S.I	Exact			B.S.I.	Exact			B.S.I.	Exact			B.S.I.	Exact
	in.	mm	mm		in.	mm	mm		in.	mm	mm		in.	mm	mm
80	0.0135	0.35	0.34	50	0.0700	1.80	1.78	20	0.1610	4.10	4.09	K	0.2810	9/32in.	7.14
79	0.0145	0.38	0.37	49	0.0730	1.85	1.85	19	0.1660	4.20	4.22	L	0.2900	7.40	7.37
78	0.0160	0.40	0.41	48	0.0760	1.95	1.93	18	0.1695	4.30	4.30	M	0.02952950	7.50	7.49
77	0.0180	0.45	0.46	47	0.0785	2.00	1.99	17	0.1730	4.40	4.39	N	0.3020	7.70	7.67
76	0.0200	0.50	0.51	46	0.0810	2.05	2.06	16	0.1770	4.50	4.50	0	0.3160	8.00	8.03
75	0.0210	0.52	0.53	45	0.0820	2.10	2.08	15	0.1800	4.60	4.57	P	0.3230	8.20	8.20
74	0.0225	0.58	0.57	44	0.0860	2.20	2.18	14	0.1820	4.60	4.62	Q	0.3200	8.40	8.43
73	0.0240	0.60	0.61	43	0.0890	2.25	2.26	13	0.1850	4.70	4.70	R	0.3390	8.60	8.61
72	0.0250	0.65	0.64	42	0.0935	3/32in.	2.37	12	0.1890	4.80	4.80	S	0.3480	8.80	8.84
71	0.0260	0.65	0.66	41	0.0960	2.45	2.44	11	0.1910	4.90	4.85	T	0.3580	9.10	9.09
70	0.0280	0.70	0.71	40	0.0980	2.50	2.49	10	0.1935	4.90	4.92	U	0.3680	9.30	9.34
69	0.0292	0.75	0.74	39	0.0995	2.55	2.53	9	0.1960	5.00	4.98	V	0.3770	3/8in	9.58
68	0.0310	1/32in.	0.79	38	0.1015	2.60	2.58	8	0.1990	5.10	5.06	W	0.3860	9.80	9.80
67	0.0320	0.82	0.81	37	0.1045	2.65	2.64	7	0.2010	5.10	5.11	X	0.3970	10.10	10.081
66	0.0330	0.85	0.84	36	0.1065	2.70	2.71	6	0.2040	5.20	5.18	Y	0.4040	10.30	0.26
65	0.0350	0.90	0.89	35	0.1100	2.80	2.79	5	0.2055	5.20	5.22	Z	0.4130	10.50	10.49
64	0.0360	0.92	0.91	34	0.1110	2.80	2.82	4	0.2090	5.30	5.31				
63	0.0370	0.95	0.94	33	0.1130	2.85	2.87	3	0.2130	5.40	5.41				
62	0.0380	0.98	0.97	32	0.1160	2.95	2.95	2	0.2210	5.60	5.61				
61	0.0390	1.00	0.99	31	0.1200	3.00	3.05	1	0.2280	5.80	5.79				
60	0.0400	1.00	1.02	30	0.1285	3.30	3.26	A	0.2340	15/64in.	5.94				
59	0.0410	1.05	1.04	29	0.1360	3.50	3.45	B	0.2380	6.00	6.04				
58	0.0420	1.05	1.07	28	0.1405	9/64in.	3.57	C	0.2420	6.10	6.15				
57	0.0430	1.10	1.09	27	0.1440	3.70	3.66	D	0.2460	6.20	6.25				
56	0.0465	3/64in.	1.18	26	0.1470	3.70	3.73	E	0.2500	1/4in.	6.35				
55	0.0520	1.30	1.32	25	0.1495	3.80	3.80	F	0.2570	6.50	6.53				
54	0.0550	1.40	1.40	24	0.1520	3.90	3.86	G	0.2610	6.60	6.63				
53	0.0595	1.5050	1.51	23	0.1540	3.90	3.91	H	0.2660	7/64in.	6.75				
52	0.0635	1.60	1.61	22	0.1570	4.00	3.99	I	0.2720	6.90	6.90				
51	0.0670	1.70	1.70	21	0.1590	4.00	4.04	J	0.2770	7.00	7.03				

2) BA Sizes and Tapping Drills

Thread	TPI	Major Dia.	Pitch	Tapping Drill	
0	25.4	6.0mm	1.00mm	No. 5	5.1mm
1	28.2	5.3mm	0.9mm	No. 14	4.5mm
2	31.3	4.7mm	0.81mm	No. 22	4.0mm
3	34.8	4.1mm	0.73mm	No. 29	3.45mm
4	38.5	3.60mm	0.66mm	No. 31	3.00mm
5	43.1	3.20mm	0.59mm	No. 36	2.65mm
6	47.9	2.8mm	0.53mm	No. 41	2.30mm
7	52.9	2.5mm	0.48mm	No. 45	2.05mm
8	59.1	2.2mm	0.43mm	No. 49	1.80mm
9	65.1	1.9mm	0.39mm	No. 52	1.55mm
10	72.6	1.7mm	0.35mm	No. 54	1.40mm
11	81.9	1.5mm	0.31mm	3/64in.	1.20mm
12	90.7	1.3mm	0.28mm	No. 57	1.05mm
13	102.0	1.2mm	0.25mm	No. 60	0.975mm
14	110.0	1.0mm	0.23mm	No. 66	0.775mm

Bibliography

This is by no means complete but covers most of the needs you may encounter in your early loco-building career.

Aster Hobbies, *The Aster Manual and Catalogue of Live Steam Engines* (from Aster Hobbies, PO Box 61, Abbot's Langley, WD5 0ZJ)

Davidson, Glenn D., *Tool Sharpening and Grinding Handbook* (Blandford Press)

Gauge One Model Railway Association, *The Project: An LMS 4f Design for G1 and other designs* (The Gauge One Model Railway Association)

Jones, Peter, *Dacre Booklet and Drawings* by Peter Jones (Brandbright Ltd)

Jones, Peter, *Practical Garden Railways* (The Crowood Press)

Rosling Bennett, A., *The Chronicles of Boulton's Siding* (David and Charles)

Sharman, Mike, *Boulton's Sidings: A Selection of Locomotive Drawings* (Oakwood Press)

Wear, Russell and Lees, Eric, *Stephen Lewin and the Poole Foundry* (Industrial Railway Society)

Any writings by 'LBSC'. Although of considerable antiquity and mostly aimed at engineering in larger scales, they make for encouraging reading. His book on 'Tich' is particularly friendly and helpful.

There are many books to be had on model engineering, but for the beginner I recommend:

Bray, Stan, *Metalworking Tools and Techniques* (The Crowood Press)

Bray, Stan, *Making Simple Model Steam Engines* (The Crowood Press) (concentrating on stationary engines but invaluable)

Bray, Stan, *Milling* (The Crowood Press)

Specialist Interest Model Books Ltd

Provides an extensive range of modestly priced small books about aspects of engineering, such as soldering and brazing, sheet metalwork, all aspects of lathe operation and so on. The book on drills, taps and dies contains extensive conversion tables. Highly recommended.

P.O. Box 327, Poole, Dorset, BH15 2RG
www.specialinterestmodelbooks.co.uk

Keith Bucklitch offers a range of excellent drawings and information for 16mm live steam locomotives that are entirely scratchbuilt.

1. A small 'quarry' Hunslet 0-4-0ST loco: 'Wild Rose Too'. This is a generic loco, with inside steam chests and slip eccentric valve gear. Methylated spirit firing.
2. A larger Hunslet 0-4-0ST loco: Charles, as preserved in Penrhyn Castle museum. Laser-cut frames and brass etch for cab and watertank are available. Slip eccentric valve gear. Meths or gas firing.
3. A Kerr-Stuart 0-4-2ST loco: Brazil. The prototype for this runs on the Sittingbourne and Kemsley railway in Kent. Hackworth valve gear. Meths/gas firing. Laser-cut frames are available for this engine.
4. A Bagnall 0-6-0T loco – Dennis – from the Snailbeach Railway. Bagnall-type valve gear. Gas firing.

No castings are required for any of these engines; standard sizes of stock material are all that is needed.

Drawings may be purchased from:

Keith Bucklitch, 308 Birmingham Rd, Bromsgrove, Worcs, B61 0HJ.
Email: keith.bucklitch@which.net for further details.

Check current editions of the magazines, *Model Engineer* and *Engineering in Miniature* for an extensive list of suppliers of engineering tools, materials and drawings. Also in the UK: *Garden Rail* (by Atlantic Press).

In Germany, seek *Gartenbahn* for resources.

In the USA: *Steam in the Garden* has a content that reflects its title; *Garden Railways* is a superb general magazine for garden railways; *Live Steam* magazine can be valuable, but has a much broader remit than the above two titles.

The Internet offers a huge resource for information. The scale groups in particular often have their own discussion sites and forums.

Useful Addresses

SUPPLIERS

The list is by no means complete, but it notes some of the more useful suppliers. It is correct at the time of publication. Web addresses in particular are prone to changes. However, there are always extensive links across the hobby.

Aster Hobbies (UK) LLP
PO Box 61, Abbots Langley, Herts WD5 0ZJ
Tel: 01932 269662
www.asterhobbies.co.uk
(Complete locomotive kits)

Blackgates Engineering
Unit 1 Victory Court, Flagship Square, Shaw Cross Business Park, Dewsbury, WF12 7TH

Brandbright Ltd
The Old School, Cromer Road, Bodham, Nr Holt, Norfolk NR25 6QG
Tel: 01263 588755
www.brandbright.co.uk
(16mm steam locomotives, chassis, boilers, fittings)

Ian Cherry
21 Norfolk Avenue, Thornton, Cleveleys, Lancs FY5 2DX
Tel/fax: 01253 852897
Email: ian.cherry@virgin.net
(Painting and lining of 0g and G1 locomotives)

Chronos Engineering Supplies
Unit 14, Dukeminster Estate, Church Street, Dunstable, LU5 4HU
Tel: 01582 471900
www.chronos.ltd.uk
(Engineering supplies and tools)

CUP Alloys
15 Sandstone Avenue, Walton, Chesterfield, S42 7NS
Tel: 01246 710892
www.cupalloys.co.uk
(Brazing and soldering materials and information)

DJB Engineering Ltd
17 Meadow Way, Bracknell, Berks RG42 1UE
Tel: 01344 423256
Email: davidbailey8@ntlworld.com
(Part-built locos, components, ready-made coal-fired boilers)

EKP Supplies
The Old Workshop, Bratton Fleming, Barnstaple, EX31 4SA
Tel: 01598 710892
Email: ekpsupplies@btinternet.com
(Engineering supplies)

Griffiths Engineering
Tel: 01298 871633
www.lathes.co.uk
(Lathes and lathe information)

HB Engineering
12 Grove Park, Tring, Herts HP23 5JL
Email: habco@tiscali.co.uk
(Gauge 1 locomotive drawings)

Hobbies Dereham (1895) Ltd
34–36 Swaffham Road, Dereham, Norfolk NR19 2QZ
Tel: 01362 692985
www.alwayshobbies.com
(Spraying and painting equipment, model-making tools)

LightLine
28 Marten Drive, Netherton, Huddersfield, West Yorks HD4 7JX
Tel: 01484 316007
www.lightline16mm.com
(Lining and lettering service for G scale and 16mm locomotives)

Machine Mart, UK chain
www.machinemart.co.uk
(Engineering machine and hand tools)

Ron M. Grant
17 Uplands Way, Springwell Village, Gateshead, NE9 7NQ
Tel: 0191 4163826
www.gratech.org.uk/rmg
(Loco name, number and worksplates, particularly 16mm/G scale)

Roundhouse Engineering Co. Ltd
Units 6–9 Churchill Business Park, Churchill Road, Wheatley, Doncaster, DN1 2TF
Tel: 01302 328035
www.roundhouse-eng.com
(16mm steam locomotive chassis, boilers, assemblies, components)

W. Hobby Ltd
Unit A, Knights Hill Square, London SE27 0HH
Tel: 020 8761 4244
www.hobby-kits.co.uk
(Useful materials, spraying equipment, modelling tools, paints)

SOCIETIES AND GROUPS

Association of 16mm Narrow Gauge Modellers
Hon Sec: Neil Bowden, 26 Cranfield, Plympton, Plymouth, PL7 4PF
www.16mm.org.uk
In particular, there is a very extensive and useful list of links to other railways, suppliers and builders at www.16mm.org.uk/16mmlink.htm

Gauge O Guild
Tel: 0845 603 6213
www.gauge0guild.com

Gauge One Model Railway Association
PO Box 581, Northampton, NN6 0YW
www.gaugeone.org

Gauge 3 Society
PO Box 7814, Sleaford, Lincolnshire, NG34 9WW
www.gauge3.co.uk

2.5in Gauge Association
Tel: 01782 396990
Email: peter@desalisjohnston.orangehome.co.uk.

There seems to be a society formed for nearly every major British railway company from the past. Because the LNWR Motor Tank is a featured design, I will give that railway society a specific mention:

L&NWR Society Membership Secretary
16 Westminster Green, Handbridge, Chester, CH4 7LE
Tel: 01244 675824
www.lnwrs.org.uk

Index